PHYSICS OF THE EARTH AND PLANETS

PHYSICS OF THE EARTH AND PLANETS

A. H. COOK, F.R.S.

Jacksonian Professor of Natural Philosophy
University of Cambridge

A HALSTED PRESS BOOK

JOHN WILEY & SONS

NEW YORK—TORONTO

© A. H. Cook 1973

First published 1973 by
THE MACMILLAN PRESS LTD
London and Basingstoke
Associated companies in New York Melbourne
Dublin Johannesburg and Madras

Published in the U.S.A. and Canada by Halsted Press
a Division of John Wiley & Sons, Inc.
New York

Library of Congress Cataloging in Publication Data
Cook, Alan H.
Physics of the earth and planets.

"A Halsted Press book."
1. Geophysics. 2. Planets. I. Title.
QC806.C65 1973   551   72-12261
ISBN 0-470-16910-9

Printed in Great Britain

Contents

Preface

This book is intended for undergraduates taking geophysics as a main or subsidiary subject and as background reading for postgraduate students of the Earth sciences. My aim is to give an account of the methods of geophysics and of the picture of the solid Earth that has been built up in the last fifty years or so; in so doing it is not possible now, if it ever was, to leave out the Moon and the planets. I have taken as a starting point the mathematical and physical knowledge of a first- or second-year physics student in a British university and, in particular, I have supposed that such readers will have a better acquaintance with the physics of atoms and solids than they will have with classical mechanics. That is one of the reasons why I have given a much fuller treatment of the gravity field, seismic waves and the dynamics of the Earth than of the topics connected with the thermal and electrical properties of the Earth; the other reason is that our knowledge of the first group is much more extensive and definite than that of the second, fascinating though the latter are and destined as they may be to develop. I have not hesitated to include discussions of some matters that are still speculative and debatable, for there lie the growing points of our study, and I have concluded with a chapter indicating how the physical study of the Earth is being extended to the planets; here again dynamical studies are of dominant importance, justifying the emphasis given them in the earlier chapters.

I have used S.I. units throughout. The S.I. system is based on seven fundamental units, the second, the metre, the kilogramme, the ampere, the degree Kelvin, the candela and the mole, and all other units are expressed in terms of those seven. In some ways S.I. units are natural ones for the study of the Earth, being based on its size (metre) and speed of rotation (second), but in other ways they may seem at first sight less convenient—the unit of magnetic field strength, the tesla, may be unfamiliar. However, the simplification and lack of confusion that come from a consistent use of S.I. units make it well worth while to adopt them to the exclusion of all others. Some of the examples have been included with the aim of helping readers to become familiar with S.I. units and with the order of magnitude of quantities associated with the Earth.

I hope this book will give its readers, in however small measure, some of the enjoyment and fascination I have found in elucidating the mysteries of the Earth and its neighbours in space.

I am grateful to Dr. W. H. Michael for permission to reproduce the material of figures 2.20 and 2.21 and to Dr. L. Essen for figure 4.8.

Symbols

A	equatorial moment of inertia
a	equatorial radius
B	magnetic induction; moment of inertia
b	polar radius
C	polar moment of inertia
C_{mn}	coefficient of gravity potential
C_p	specific heat at constant pressure
c	speed of light
D	distance
d	distance
E	energy
f	polar flattening
G	constant of gravitation
g	acceleration due to gravity
H	dynamic ellipticity $= (C - A)/C$; magnetic field
h	height; Love's number
g_n^m, h_n^m	coefficients of geomagnetic field
I	moment of inertia
J_n, J_{nm}	coefficient of gravity potential
K	bulk modulus
k	Love's number; $2\pi/\lambda$; Boltzmann's constant
l	length; Love's number
M	mass
$\mathbf{M}$	angular momentum
m	mass
n	refractive index
p	pressure
r	radius
S	dipole moment
S_{mn}	coefficient of gravity potential
T	kinetic energy; period
t	time; thickness
U	energy
V	potential energy
X	magnetic field (N component)
x	coordinate
Y	magnetic field (E component)
y	coordinate

Z	magnetic field (vertical component)
z	coordinate
α	compressional wave velocity
β	shear wave velocity
γ	unit of 10^{-9} tesla
δ	density
κ	thermal conductivity
λ	wavelength; Lamé parameter
μ	shear modulus; magnetic permeability
$\boldsymbol{\mu}$	magnetic moment
ν	frequency
ρ	radius; density
σ	Stefan's constant; density; electrical conductivity
θ	colatitude
ϕ	longitude
Φ	$\alpha^2 - \frac{4}{3}\beta^2$
ω	angular frequency

CHAPTER ONE

Introduction

'Ay me, what is this world?'

Shakespeare, *King Henry VI*, Pt. 2, III, 2

Almost from the first times of which we have record, men have stood on a hill and have gazed at the rocks below their feet and at the planets above them and have wondered at the beauty and the mystery of nature. How was the world made, how did it come into its place in the heavens, what moved the mountains, what sustains the stars? As man applied his reason to nature, he saw that the rocks on which he stood had once been below the seas or had been molten in the bowels of the Earth, and yet now are raised up high above the plains below; and he saw, as Galileo first did with his telescope, that the planets are not embedded in mysterious spheres but are like the Earth itself. So geologists came to study the history of the surface of the Earth, the seas, the mountains, the rocks and the things that had lived in the past, and astronomers came to study the planets, and how, under the sway of universal gravity, they move around the Sun, and they ask how the planets can have come into being. Now, in later times, another study has developed, the study of the physics of the whole Earth, not just the surface; and of the Earth in its own right, not just as one of the companions of the Sun. This is geophysics and it is the subject of this book.

The geophysicist does not concentrate on any particular part of the Earth but tries to see it as a whole; his main interest is the Earth, and, when he looks at the planets, it is so that he may extend his knowledge of the Earth by seeing the Earth in their light and them in its. The methods of the geophysicist are those of physics, methods that have distinguished physics from other branches of natural knowledge from the time of Galileo and Newton up to the present time—methods that are summarized and exemplified in Newton's laws of motion. Newton's laws of motion enable us to investigate the forces between bodies by examining the changes of motion that the forces produce; at the present time the methods of physics are perhaps seen at their clearest in high-energy physics where the forces between elementary particles are investigated by observing the changes in the motions of such particles when others are directed at them with the help of special accelerators. The idea of motion or change is not limited by mechanical motions and acceleration but the electrical and thermodynamical forces of nature may be questioned in the same way.

Newton's laws of motion have set the pattern of physics in another way— they are in mathematical form, and the chain of reasoning that leads from observations of the changes of motion to the forces that produce those changes is a mathematical argument. Mathematical reasoning is the most exact, most power-ful and most definite of which we are capable, and is associated with a third charac-teristic of the methods of physics (common indeed to all scientific studies but more obvious in physics than elsewhere), the abstraction of a particular group of self-contained phenomena to be studied by themselves—for example, if we are measur-ing the magnetic moment of a bar magnet, we usually believe it to be rather un-important that the magnet is painted blue instead of red.

The physicist does not attempt to give a complete account of all phenomena that present themselves to him but he concentrates attention on those that he believes to be significant. Even within a limited field, the physicist attempts to set up orders of significance, believing that if he understands certain aspects of a problem then the understanding of others will be more a matter of application of rules than of originality or insight. Most physicists would probably agree that, while there is an immense amount of detailed work still to be done in, say, atomic spectroscopy, or the physics of the solid state, the principles of quantum mechanics have established the rules upon which atoms or crystals are built and the mathe-matical tools for calculating them in detail have been sketched out.

In geophysics, the characteristic methods of physics are applied to the study of the whole Earth. We examine the Earth by seeing how it influences other bodies and how it responds to forces that act upon it. We make the observations with the techniques of physics and we interpret them in terms of the constitution of the Earth by mathematical reasoning, concentrating on what we believe to be the most significant facts about the Earth as a whole.

Broadly speaking, we rely on mechanical investigations for our knowledge of the structure of the Earth, to begin with, the way in which the Earth attracts bodies in its neighbourhood. From the value of the acceleration due to gravity at the surface of the Earth and from the motions of artificial satellites about the Earth, we can find the size, the mass and the shape of the Earth and something about the density of the materials of which different parts of the Earth are made. But we are very limited in what we can learn of the Earth, for we can make our observations only at the surface of the Earth and only at the present time, and, when we try to work out the structure of the inside of the Earth, we find that we can do so in many ways, and the laws of physics tell us that few of our interpreta-tions can be unambiguous. Gravity measurements alone can give us no detailed information about the density in the interior of the Earth and we must call upon other observations.

When the Earth is struck, it quivers: if an earthquake occurs or a nuclear bomb is exploded, waves of elastic vibrations, like sound waves in a metal bar, spread throughout the Earth and can be detected even at the antipodes if the bomb or the earthquake is great enough. The time a wave takes to travel from the source back to the surface carries information about the density and elasticity of the substance of the Earth which extends in a very detailed way the indications we obtain from gravity around the Earth. Further information comes from the way in which the Earth rings like a bell when made to do so by a sufficiently large earthquake and how it yields elastically under the gravitational attraction of the Sun and the Moon.

If all the data from gravity, earthquake waves and the elastic vibrations and yielding of the Earth as a whole are put together, we obtain a picture of the increase of density and pressure in going from the outside to the centre of the Earth, and it turns out that the Earth is divided very sharply into two zones, the outer called the *mantle*, and the inner called the *core*. The mantle is solid and made of an ionic crystalline material not unlike the crystals that form some rocks at the surface and which, under the conditions of the interior of the Earth, must be electrically a semiconductor. The core is liquid and a metal, being composed most probably of iron and nickel for the most part, and is an electrical conductor. Of course, these inferences about the materials of the core and the mantle cannot be made without the results of experiments in the laboratory on materials that might have the same mechanical properties as those in the interior of the Earth, although such experiments are difficult and limited because of the high pressures and temperatures that must be used.

The Earth is not alone in space and man has speculated that the Moon and the nearer planets may be in some ways similar to the Earth. Artificial satellites and space probes, supplementing the natural satellites that a few planets possess, have allowed some of the mechanical observations that are made about the Earth to be made about the planets and we have quite a good idea of the sizes, masses and densities of the nearer ones. It turns out that the relation between density and pressure in the Earth follows a remarkably simple rule that seems to apply almost everywhere despite the very great range of density and the different sorts of material in the Earth, and it is well worth while to suppose to start with that the same rule applies to the planets and to work out the consequences. The results of such calculations, which may be regarded as a summary in compact form of our knowledge of the mechanical properties and constitutions of the Earth and its immediate neighbours, are given at the end of chapter 5 as the conclusion to a discussion of the way in which mechanical observations enable the internal structures to be inferred.

But the Earth is changing all the time and we naturally ask what changes are occurring, how fast are they taking place, and do we know of any source of energy that is driving them? Radioactive decay supplies the time scale and the clock which we need to measure rates of change. We can now establish times in the Earth and on the Moon back to 4500 My, in some detail in the more recent past but rather sketchily in the remote past. When radioactive nuclides decay within the Earth they generate heat that may drive the changes that go on in the Earth, but the study of the flow of heat from the Earth and of the temperature within the Earth is complex, both because of the difficulty of the mathematics and because many of the data that are needed are extremely uncertain. In particular, we do not know the temperature of the Earth when it was formed and therefore how much heat it contained when it came into being, for the answer depends on our ideas about the formation of the solar system and the origin of the Earth, and so, although geophysicists try to study the Earth so far as possible without having to depend on knowledge that the astronomer has, here in this very important field, they must wait upon the astronomer's idea of how the Sun evolved and how the planets were placed around it; the geophysicist may be able to make a return to the astronomer, for it may be that some ideas about the origin of the solar system would not agree with the Earth as we find it.

Man has used the lodestone to point to the north for longer than he has realized how gravity acts and for more than three hundred years has known that the magnetic field of the Earth is changing slowly, at the rate of about ten per cent in two or three hundred years. The field undergoes even greater changes in longer periods, as shown by the permanent magnetization of baked pottery and lavas acquired by the material on cooling from a high temperature in the local magnetic field of the time. The most remarkable result is that the field has not always pointed in the same direction but has turned through 180° very many times.

The Earth is accompanied through space by the planets; very roughly they are of much the same size as the Earth, they are cold as the Earth is instead of being hot like a star, they have similar densities and some of them must surely be made of much the same material as the Earth. We are continually learning more about them as astronomical observations improve and especially as space probes are sent to the closer planets. The Moon and the planets are in some ways each quite different from the Earth and each other but there are sufficient similarities for them to tell us something about the Earth, and so the last chapter of this book is concerned with the planets and how we may learn about the Earth from them and about them from the Earth.

In geophysics we stand on the surface of the Earth and try to imagine what it is like inside. The task is very difficult for, unlike Jules Verne's traveller, we cannot voyage to the centre of the Earth. All we can say is, if the Earth were such and such inside, then we should find that its gravitational attraction, its elastic oscillations, and so on, would be just those that we observe. But many different models for the Earth may well give the same results at the surface, and many different ideas of the history of the Earth may be consistent with the state in which we now find it. We just have to accept this situation.

It is one that encourages speculation. Speculation is indeed part of scientific advance, and helps to map out possible models of the Earth, but there comes a time when speculation must be controlled by working out just how wide a range of internal conditions in the Earth would lead to what we actually observe at the surface. This can only be done mathematically and for the most definite results our data should be in the form of numbers—measurements of the attraction of gravity, numerical values of the periods of free oscillation of the Earth and so on. Then we can employ the most rigorous methods of reasoning to our data and we shall also have a measure of the reliability of our data.

It is indeed striking how most of the big advances in geophysics, as in other branches of physics, have come when someone has found a way of measuring something that could not previously be measured or of measuring more simply or more accurately something that before could only be measured inaccurately or with difficulty. Our understanding of the Earth has improved immensely in the last two decades largely as a result of four great advances in making measurements on the Earth, namely, measurements of the acceleration due to gravity at sea on a surface ship; the launching of artificial satellites; the development of magneto-meters that could be towed behind ships or aircraft or carried in satellites or space probes; and the development of refined instruments for detecting the elastic waves from nuclear explosions. Geophysics cannot be understood properly unless the methods used for making measurements and, in particular, the limitations to which they may be subject are appreciated.

I say little about the way in which the methods of geophysics and the knowledge acquired by them may be applied to geology, to prospecting and to engineering, but nonetheless geophysics is not only fundamental to our understanding of the natural world in which we live but is also essential to the proper management of natural resources and to our appreciation of the physical limitations within which we must live our lives. One of the most fascinating of these aspects of geophysics is the study of the occurrence of earthquakes, a study which is important for the safety and well-being of large populations in many parts of the world, which requires refined and elaborate methods of observation and demands a deep understanding of the changes going on in the Earth that as yet we do not possess.

CHAPTER TWO

The gravity fields of the Earth and the planets

'l punto al qual si traggon d'ogni parti i pesi.'

Dante, Inferno, XXIV

2.1 Introduction

We begin our study of the Earth with simple questions that almost anyone can answer for himself—what is the acceleration of a body falling to the Earth and how do the orbits of artificial satellites change as they circle the Earth? Everyone is so familiar with gravity which acts upon us continually that it may be difficult to realize that quite simple observations of how the Earth attracts bodies near to it can tell us a great deal about the Earth and so are the natural starting point for our journey in imagination to the centre of the Earth. Taken together with the size of the Earth, the acceleration due to gravity enables us to calculate the mass and density of the Earth and so gives us an idea of the materials of which the Earth is made, and, if we look more closely at how the acceleration varies over the surface, we may learn something of the structure of the Earth in more detail. Artificial satellites moving under the gravitational attraction of the Earth have turned out to be most effective in the study of gravity, and our knowledge has been quite transformed in the decade and a half since the first one was launched. They are at least as important in the study of the planets.

If all we know about a planet are its size and acceleration due to gravity, our ideas about its internal state are very general for we are limited to estimates of its mass, its density and possibly how strongly it is concentrated towards the centre, together with an indication of variations of density close to the surface. If we are to gain more detailed knowledge our study of gravity must be accompanied by other investigations; our study of gravity serves then as an introduction to the study of the mechanical properties of the Earth but it is all we have to go on in investigating the planets.

If two point masses m_1 and m_2 are at a distance r apart, they attract each other with a force equal to $GM_1 m_2/r^2$. If the masses are given in kilogrammes, the distance in metres and the force in newtons, the Newtonian constant G is $6·67 \times 10^{-11}$ N m^2 kg^{-2} (the units of G might also be written as m^3 kg^{-1} s^{-2}). Since acceleration equals force divided by mass, mass 1 has an acceleration Gm_2/r^2 towards mass 2, while mass 2 has an acceleration Gm_1/r^2 towards m_1, each being expressed in metres per second per second. The force acting on the mass m_1 is equal to

$$-m_1 \frac{d}{dr}\left(-\frac{Gm_2}{r}\right)$$

and the force acting on m_1 in a direction x due to the attraction of a set of other masses at distances r_i from m_1 can be written as

$$-m_1 \frac{\partial}{\partial x}\sum_i -\frac{Gm_i}{r_i}$$

These well-known results lead to the idea of the potential. The gravitational force on m_1 due to any distribution of masses is $-m_1 \operatorname{grad} V$, where V, the potential, is equal to $\int -G\,dm/r$, dm being the element of mass at the point with position vector $\mathbf{r}$. The acceleration of m_1 is independent of its own mass, and so in dealing with the gravitational attraction of the Earth it is convenient to take g to be the acceleration, that is, m_1 is understood to be 1 kg. Thus,

$$\boldsymbol{g} = -\operatorname{grad} V \tag{2.1}$$

It is a direct consequence of the inverse square law that V satisfies Laplace's equation, $\nabla^2 V = 0$, in empty space while, if the local density of mass is ρ, it satisfies Poisson's equation, $\nabla^2 V = 4\pi\rho$.

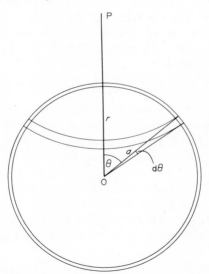

Figure 2.1 The attraction of a spherical shell

As an elementary illustration of the utility of the potential, consider the attraction of a spherical shell (figure 2.1). Let a be the radius of the shell and let the mass of the shell per unit area be σ. Let the potential be calculated at a point P at a distance r from the centre O of the shell. Consider the strip of the shell cut off by the two cones of semi-angles θ and $\theta + d\theta$ with centres at O and with OP as common axis. The area of the strip is $2\pi a^2 \sin\theta \, d\theta$ and the mass is $2\pi\sigma a^2 \sin\theta \, d\theta$. Every point on the strip is at a distance $(r^2 + a^2 - 2ar\cos\theta)^{1/2}$ from P and so the potential of the strip at P is

$$-G2\pi\sigma a^2 \sin\theta \; d\theta / (r^2 + a^2 - 2ar\cos\theta)^{1/2}$$

The potential of the complete shell is obtained by integrating this expression from 0 to π, that is

$$V = -2\pi G\sigma \frac{a}{r}\{(r + a) - (r - a)\}$$

or

$$V = -4\pi G\sigma a^2/r \tag{2.2}$$

But the mass of the shell is $4\pi a^2 \sigma$ and so the potential of the shell is the same as that of an equal mass concentrated at the centre. It follows that the acceleration due to the attraction of a shell of mass M is GM/r^2 towards the centre of the shell.

Now suppose that a sphere is composed of a set of spherical shells so that the density of the sphere may vary with radius but does not depend on orientation. Adding up the potentials of the separate shells into which the sphere may be supposed to be divided, it will be seen that the potential of the whole sphere is $-GM/r$, where M is now the mass of the sphere. When Newton obtained this result he saw that so far as their mutual attraction was concerned the Earth and the Moon could be replaced by point masses, and so he could complete the calculations which convinced him of the universal validity of the inverse square law of gravitational attraction.

The simple result for the potential of a sphere with any radial distribution of density is important because it shows that the only fact that can be learnt about a spherically symmetrical body from measurements of gravity is the mass of the sphere; so long as the density within a sphere depends only on the distance from the centre nothing else can be learnt. It is only because the Earth is not quite spherically symmetrical that gravity measurements tell us anything about the distribution of density with radius. Venus, on the other hand, is as nearly as can be measured at present, quite symmetrical about her centre, and gravity measurements will never tell us anything about her internal state apart from her mean density.

In the S.I. system, acceleration is measured in metres per second per second, and the value of the acceleration due to gravity at the surface of the Earth is just under 10 m s^{-2}. The variations around the Earth are much less. The polar flattening corresponds to an increase of 0·05 m s^{-2} from equator to poles, or about 10^{-5} m s^{-2} for each kilometre travelled from south to north in mid-latitudes. The acceleration decreases with height, at a rate of between 2 and 3 $\times$ 10^{-6} m s^{-2} per metre, while changes in the density of rocks beneath the surface may give rise to variations of up to 10^{-3} m s^{-2}. The variations are thus much less than the value of the acceleration itself and it is convenient to have a much smaller unit than 1 m s^{-2}; a convenient and generally used unit is the milligal (mgal) equal to 10^{-5} m s^{-2} or one part in a million of the acceleration due to gravity.

2.2 The Measurement of Gravity

The acceleration due to gravity measured at the surface of the Earth is the acceleration of a body falling freely and is normally measured in the frame of reference of a laboratory rotating with the surface of the Earth. The measured acceleration is therefore not just that of the gravitational attraction of the material of the Earth at the point of measurement but is the resultant of that acceleration and the acceleration corresponding to the rotation of the laboratory about the polar axis of the Earth. Values of acceleration are always given in the rotating frame, but it must be remembered, when comparing them with satellite data, that the latter are given in a non-rotating frame. The details of the comparisons will be given later.

Since the range of acceleration due to gravity over the Earth does not exceed 1 part in 200 and since the variations of geological and geophysical interest are considerably less, the problems of measurement divide clearly into two—the measurement of the acceleration absolutely and the measurement of differences from place to place. The absolute measurement, that is the measurement of an acceleration in terms of the units of the metre and the second, is needed to find the size and the mass of the Earth, as well as in metrological physics, and until recently it has not been possible to make absolute measurements with the precision of measurements of differences. On the other hand, many geophysical studies can be made without any exact knowledge of the absolute value and so quite different techniques are used for the absolute measurement and for the measurement of differences.

Absolute measurements are now made by the direct observation of the free fall of some object, the distance over which the object falls being measured by optical interference while the times are measured electronically. Diagrams of the apparatus used in three recent determinations are shown in figures 2.2, 2.3 and 2.4. In the determination at the National Physical Laboratory, a glass ball was thrown vertically upwards and signals were obtained from photomultipliers as it passed pairs of slits and focused one slit of a pair upon the other. The electrical signals were used to start and stop electrical timing systems that measured the times spent by the ball above each of the pairs of slits. The slits were mounted upon glass blocks, the vertical separation of which was measured by an interferometer. The whole experiment was performed in vacuum to eliminate the errors due to the resistance and buoyancy of the air. Let T_1 be the time spent by the ball between its upward and downward passages across the lower pair of slits and let T_2 be the corresponding time spent above the upper pair of slits.

Let H be the vertical separation of the slits. Then it follows simply from the equations of motion for free fall under gravity that (Cook, 1969)

$$g = \frac{8H}{T_1^2 - T_2^2} \tag{2.3}$$

H is conveniently about 1 m and T about 0·9 s; gravity changes by about 0·3 p/M* over a height of 1 m and so to achieve an accuracy of 0·1 p/M at a particular height a small correction must be applied for the variation of gravity with height. There is no difficulty in making the measurements of length and height to better than 1 p/M but great care must be taken to eliminate systematic errors and especially the effects of any movement of the apparatus caused by throwing the ball up.

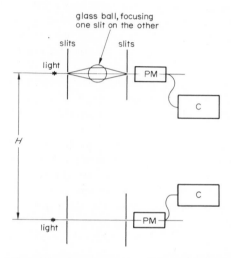

Figure 2.2 Absolute gravity determination at the National Physical Laboratory (Cook, 1967): C, counter; PM, photomultiplier

A similar, but much more accurate, experiment has been performed at the Bureau International des Poids et Mesures at Sèvres (figure 2.3). The particular features of this determination are, firstly, that the object thrown up is a cube reflector that forms part of the interferometer system, and, secondly, that the whole interferometer system is mounted on springs that cut out the effect of ground movement. The daily variations of gravity due to the tides (chapter 4) have been measured with this apparatus.

Throwing up an object is mechanically more complicated than just letting it fall, but has two advantages. One possible source of systematic errors in a free-fall measurement of gravity is that there may be timing errors that depend on the

* 1 p/M is 1 part in a million.

speed with which the object is moving, so that the error when the object is near the top of its flight is different from the error when the object is near the bottom. In a simple free-fall experiment the time intervals are measured between events occurring at different speeds and accordingly the intervals may be subject to a systematic error, whereas in the up-and-down experiment the time intervals are measured between events that occur at the same speed and that possible source of error is very greatly reduced. A further advantage of the up-and-down experiment is that the drag of the air has no effect on the result, whereas it does affect the free-fall experiment. Nevertheless, with careful design of the experiment, a good determination by the free-fall method can be made and a diagram of one experiment is shown in figure 2.4. Here the falling object is a lens mirror

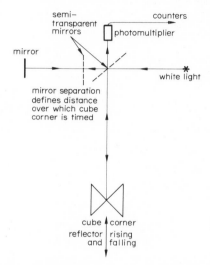

Figure 2.3 Absolute gravity determination at the International Bureau of Weights Measures (Sakuma, 1972)

reflector forming part of an interferometer. The light source is a gas laser and the interference fringes are counted electronically as the reflector falls. The reflector falls in a highly evacuated tube so that air drag is negligible and great care is taken to release it so that it does not make the apparatus vibrate, while the fixed part of the interferometer is placed on an anti-vibration mount to cut out the ground movement. This experiment again gives results accurate to 0·1 p/M.

Measurements with the free-fall apparatus, which is more easily taken from place to place than the other two, show that the three methods give results that agree to within 0·1 p/M, agreement that is better than can be obtained in pendulum measurements of the differences of accelerations due to gravity. The main reason for the attention given to the very accurate measurement of the acceleration is that it needs to be well known to establish the units of force, pressure and electrical current, all of which depend on knowing the force exerted by gravity

upon a given mass. At the same time the results establish a sound basis for calculating the mean value of the acceleration over the Earth in absolute terms and they
also provide a reliable connection between values in Europe and in North America.

The variation of acceleration over the Earth is measured by means quite different from those for the absolute value at a single site. Because the variations
are small in relation to the acceleration itself, the methods must be very accurate
and because the accelerations vary quite sharply from place to place the apparatus
should be quick to use, enabling an area to be covered speedily with many measurements. The requirements are satisfied by spring balance gravity meters in which
the gravitational force on a mass is balanced by the force exerted by a stretched
spring. There are two main problems in the construction of a gravity meter, that

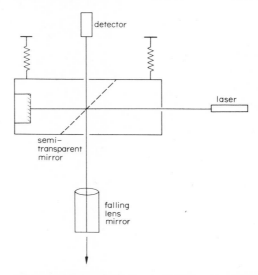

Figure 2.4 Absolute gravity determination by Hammond and Faller (1972)

of devising a means of measuring the extension of the spring sensitive to 0·01 p/M,
and that of ensuring that the gravity meter reading does not depend on factors
other than gravity. Two basic schemes have given good sensitivity. The first is
indicated in figure 2.5. A mass is carried at the end of a horizontal arm attached
to a horizontal helical spring. If the torque exerted by the spring is $\tau\theta$, where θ
is the angle through which the spring is twisted, and if l is the length of the arm
to which the mass m is attached, then

$$\tau\theta = mlg$$

Hence

$$d\theta = \theta \, dg/g$$

and, by arranging the spring constant so that θ is a large multiple of 2π the angular
sensitivity of the system can be made high. The deflection of the arm can be measured photoelectrically, and the arm may be brought to a standard position by

an auxiliary spring; the auxiliary spring is adjusted with a micrometer, the reading of which indicates the value of the acceleration due to gravity.

The other gravity meter system depends on the properties of a coiled spring which is wound so that the force F that it exerts is proportional to the extended length l of the spring, that is

$$F = kl$$

instead of the more general relation

$$F = k(l - l_0)$$

obeyed by ordinary springs. The spring is called a *zero-length spring* because l_0 is made zero.

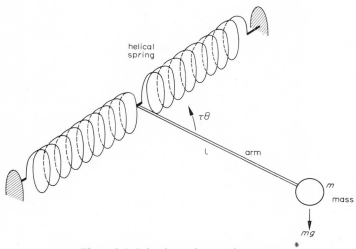

helical
spring

$\tau\theta$

l arm

m

mass

mg

Figure 2.5 Askania torsion gravity meter

When a helical spring is extended, the wire composing it is twisted, the component of the twist of all the coils along the axis of the helix being equal to the extension of the spring as a whole. By giving the wire an additional twist as it is wound into the helix, it is possible to arrange for the law of force to be very close to $F = kl$ over some range of the length l.

If an arm carrying a mass m is supported by a zero-length spring as shown in figures 2.6 and 2.7, the position of the arm can be made unstable by suitable adjustment of the points of support, and then the slightest change in the gravitational force on the mass would produce a large and arbitrary change in the position of the arm. A truly unstable system is unmanageable, but the angular position of the beam can be arranged to be very sensitive to changes in the acceleration due to gravity without being actually unstable.

In a practical instrument, the deflection of the beam is read by a sensitive

microscope or photoelectric system, and the beam is returned to a standard
position by an additional delicate spring controlled by a micrometer screw.

The stability of the reading of a spring balance gravity meter is determined by
the properties of the spring. The force exerted by a coiled spring is proportional
to the modulus of rigidity, the temperature coefficient of which is about 10^{-4}
$\deg K^{-1}$ for most materials. Special steels have been made with very much smaller
coefficients but are apt to creep with time. Fused silica springs are more stable,
but have the usual large coefficient of rigidity; the very widely used Worden

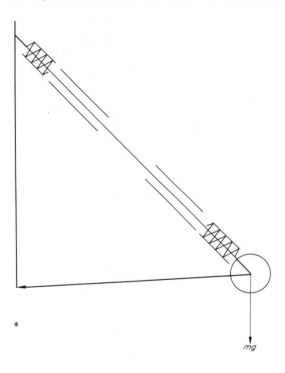

Figure 2.6 La-Coste–Romberg gravity meter

gravity meter uses a spring of fused silica with a temperature-compensating
device. The technique of the fabrication of springs has now been so greatly refined
that gravity meters stable to better than 0·1 p/M of the acceleration due to gravity
over many days are available. Although the meter must be placed in a thermostat
enclosure, the whole system can be made very light and portable and is very
quick to use. In fact the measurement of the acceleration due to gravity is now
the quickest part of a gravity survey. For meaningful measurements, the latitude
and height of the places at which they are made must be known with accuracies
corresponding to better than 0·1 mgal, or 100 m in position and 0·2 m in height,
and even in countries that have as good maps as Britain it takes longer to measure

the height and position of the place of observation with sufficient accuracy than it does to make the gravity meter readings.

The reading of a gravity meter is not a value of acceleration but some other quantity, such as the reading of a micrometer screw, that must be calibrated before it can be converted into differences of acceleration. Experience has proved that the only satisfactory way of calibrating gravity meters is to find the readings at places where the acceleration is already known. Sites of absolute measurements are of course such places, but they are not numerous enough nor satisfactorily distributed. Because the acceleration varies strongly with latitude and only irregularly with longitude, calibration sites should be arranged along north–south lines, and they should be spaced so that the differences of acceleration correspond to a large fraction of the range of the meter. The necessary reference measurements at the calibration sites are made with pendulums.

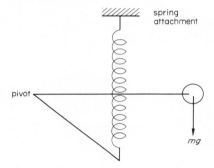

Figure 2.7 Worden fused silica gravity meter

In contrast with the considerable advances that have been made in the absolute determination of gravity over the last ten years or so, the method for the measurement of differences of acceleration by means of pendulums remains essentially that established by Captain Henry Kater (1819) more than 150 years ago. The period of a pendulum is given by

$$T = 2\pi \left(\frac{I}{mgh} \right)^{1/2}$$

so that, if the moment of inertia, I, and the distance, h, of the centre of mass from the point of support remain constant, the periods T_1 and T_2 at places where the values of acceleration are g_1 and g_2 stand in the relation:

$$\frac{T_1^2}{T_2^2} = \frac{g_2}{g_1}$$

It is a simple matter to measure the periods with high accuracy, for example by reflecting light from a mirror on the pendulum to a photocell and counting electrically the number of oscillations in a given interval (figure 2.8). It is much more difficult to ensure that the pendulum remains mechanically the same when it is carried from place to place. The big problem with pendulums is that the forces exerted by the support are far from constant. Pendulums for the measurement of gravity (figure 2.9) are supported by knife edges rigidly fixed to the pendulum and resting on a highly polished flat surface and, although to the naked eye it might seem that the knife edge is a cylinder and that it rests on a truly plane surface, yet microscopically both surfaces are very irregular and make contact at relatively few points where the pressure is very high. Thus it is not surprising that, as the apparatus is carried from place to place, the forces between the knife and the plane

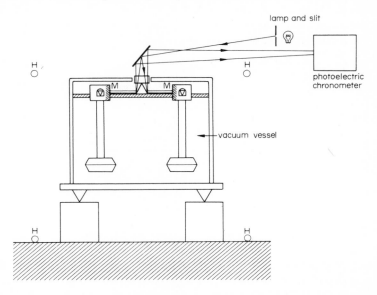

Figure 2.8 Two-pendulum gravity apparatus: H, Helmholtz coils to annual vertical magnetic field of Earth; M, mirror on pendulum

do not remain the same and, as a result, measurements of differences of acceleration with pendulums cannot be relied on to much better than 0·5 p/M even though the measurements of period at one plane would suggest uncertainties much less than 0·1 p/M. Pendulum measurements are nonetheless the basis of differential measurements over the surface of the Earth, and will remain so until sufficient well-distributed absolute measurements have been made. Pendulum measurements give differences in absolute units once the value at one point is known but the precision of which they are capable is much less than that of spring balance gravity meters, and in consequence the best calibration of the gravity meters is obtained if pendulum observations are made at the greatest convenient intervals, usually now about 500 mgal. An international network of suitable pendulum observations has been established over the past 20 years, as shown in figure 2.10.

Measurements of the acceleration due to gravity at sea are especially difficult. In the first place, a survey ship will be away from its base for weeks or months at a time so that the spring of a spring balance gravity meter has to be much less subject to creep or to sudden change than is that of a land instrument which can be checked every day or even more frequently. Secondly, the meter must respond linearly to changes of acceleration over a very wide range, on account of the variations of acceleration with the motions of the ship. Suppose that the meter reading is not simply proportional to acceleration but depends also on the square of the acceleration:

$$R = R_0 + ag + bg^2 + \ldots$$

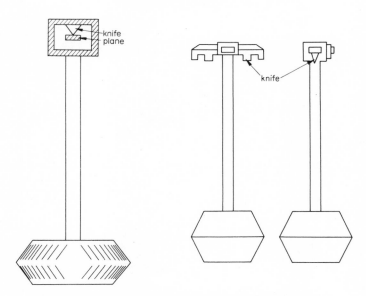

Figure 2.9 Some designs of pendulum for gravity measurement

The instantaneous acceleration is equal to the steady value, g_0, together with a periodic acceleration of the ship, $\gamma \cos \omega t$ say. Thus

$$R = R_0 + a(g_0 + \gamma \cos \omega t) + b(g_0^2 + 2g_0 \gamma \cos \omega t + \gamma^2 \cos^2 \omega t) + \ldots$$

and the mean reading averaged over a number of periods of the motion of the ship will be

$$R_0 + ag_0 + b(g_0^2 + \tfrac{1}{2}\gamma^2)$$

γ may reach as much as $0 \cdot 1\,g$, and, although a meter would not be used under such rough conditions, yet it is clear that the response has to be linear to a very high degree. Meters are now available that are stable and linear to about 1 p/M (see Worzel and Harrison, 1966).

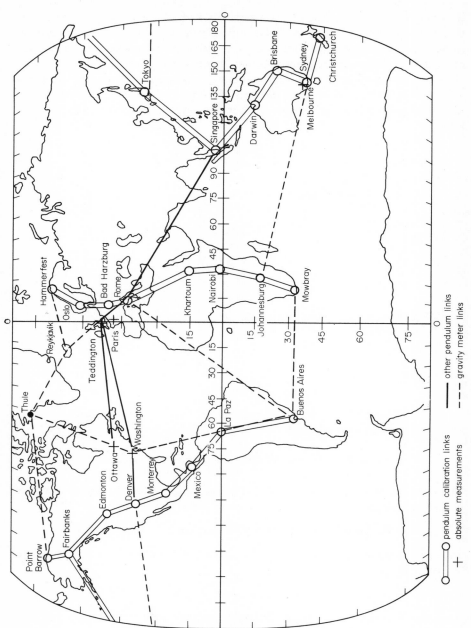

Figure 2.10 The international gravity network

If a gravity meter is freely supported so that it always records the resultant acceleration acting upon it, the apparent value of gravity will be greater than the true value on account of the horizontal accelerations of the ship. The situation is shown in figure 2.11. The acceleration in the vertical direction is $g + \ddot{z}$, where $\ddot{z}$ is the vertical acceleration of the ship. The acceleration in the horizontal direction is that of the ship, $\ddot{x}$, and so the resultant acceleration of the gravity meter is

$$\{(g + \ddot{z})^2 + \ddot{x}^2\}^{1/2}$$

or

$$g + \ddot{z} + \tfrac{1}{2}\ddot{x}^2/g$$

the mean value of which is $g + \tfrac{1}{2}\langle \ddot{x}^2 \rangle_{av}/g$. If a gravity meter is allowed to align itself along the resultant acceleration, the horizontal acceleration of the ship must be recorded so that the necessary correction may be applied to the reading of the meter. The accelerations are of course periodic with zero mean value so that they

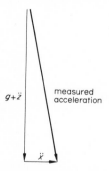

$g + \ddot{z}$ measured acceleration

$\ddot{x}$

Figure 2.11 Principle of gravity measurement at sea

can be measured by an electromechanical accelerometer of suitable free period. The correction is usually computed and applied automatically.

The alternative scheme is to place the gravity meter on a table which is controlled from a gyroscope so that it is always horizontal, that is, it is perpendicular to the mean direction of the acceleration. The measured acceleration is then

$$g + \ddot{z}$$

of which the mean value is g. Successful measurements of gravity at sea are made both with the free meter supplied with corrections for the horizontal accelerations and by the meter on a stabilized platform.

While the problems of the measurement of the horizontal acceleration of a ship at sea are severe, the largest uncertainty in the results usually comes from ignorance of the angular velocity of the ship about the polar axis of the Earth. It has already been said that the acceleration due to gravity at the surface is always measured and compared in the frame of reference rotating with the Earth, but measurements on a ship moving with respect to the surface of the Earth are not made in that standard frame. The situation is indicated in figure 2.12. If a is the radius of the Earth (considered as a sphere), if $\tilde{\omega}$ is the spin angular velocity of the Earth, v is the linear velocity of the ship from west to east relative to the surface of the Earth and g is the Newtonian attraction of the mass of the Earth, then the acceleration of the ship in latitude θ has the components g towards the centre of the Earth and $(\tilde{\omega} + v/a)^2 a\cos\theta$ perpendicular to the polar axis of the Earth.

Because the rotational acceleration is small compared with g, it is sufficient to resolve it along the radius. The radial component is

$$(\tilde{\omega} + v/a)^2 \, a \cos^2 \theta$$

which differs by $2\tilde{\omega}v\cos^2\theta$ from the value $\tilde{\omega}^2 a\cos^2\theta$ for a point rotating with the surface. The effect of the ship's surface velocity is large, amounting to about

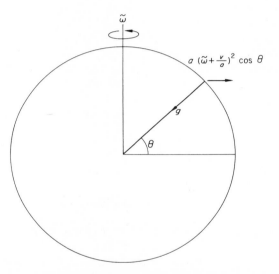

Figure 2.12 Principle of the Eötvös effect

7 mgal per knot east or west at the equator, so that for measurements to be accurate to 1 mgal the speed of the ship must be known to a few tenths of a knot in most parts of the world.

Such accuracy of measurement of the surface speed of a ship is out of the question when the ship is out of range of land. The speed relative to the water can be measured reasonably accurately but in many parts of the oceans the currents are quite unknown to the precision required and measured values of the acceleration due to gravity at sea cannot be relied on to better than 5 mgal as compared with 0·5 mgal for most land measurements. Better results must await better navigational methods in the deep oceans and, in particular, the exploitation of the possibilities of artificial satellites.

2.3 Gravity anomalies

The observed variation of the acceleration due to gravity is in part accounted for by well-understood causes and attention is therefore usually concentrated on the residual variations due to less well-known differences of density within the Earth. Such residuals are known as *gravity anomalies*.

The Earth is very close to being a rotating flattened sphere and, as will be seen at the end of this chapter, the acceleration due to gravity at sea level (g_0) over such an ideal Earth can be calculated exactly. The theory shows (p. 36) that the acceleration should increase from the equator to the poles according to the formula

$$g_0 = g_e(1 + \beta_1 \sin^2 \phi + \beta \sin^2 2\phi)^2 \qquad (2.4)$$

ϕ is the latitude and g_e is the acceleration due to gravity at the equator. The coefficient β_1 is about 5×10^{-3} and β_1 and β_2 are best found from satellite data, as shown later in this chapter.

The acceleration also decreases with height above sea level. If the Earth were a sphere of radius a, the value at a height h above the surface would simply be given by the inverse square law formula

$$g_h = g_0 \left(\frac{a}{a+h}\right)^2 = g_0 \left(1 - \frac{2h}{a}\right) \qquad (2.5)$$

provided h is much less than a. With a equal to about 6400 km, $2h/a$ is about 0·3 mgal m^{-1}.

Although the Earth is not exactly a sphere, the formula is in fact good enough for most purposes. The simplest formula for calculating the value of acceleration at latitude ϕ and height h is therefore

$$g(\phi, h) = g_e(1 + \beta_1 \sin^2 \phi + \beta_2 \sin^2 2\phi)\left(1 - \frac{2h}{a}\right) \qquad (2.6)$$

and the differences $g_{obs} - g(\phi, h)$ between the observed acceleration, g_{obs}, and the values from the formula are known as *free-air anomalies* because they are the differences between the actual values and those calculated on the assumption that there is no matter between sea level and the point of observation.

That assumption is of course usually quite unrealistic, for almost all observations that are not made at sea level are made on the surface of the ground with rock between the point of observation and sea level. The gravitational attraction of those rocks is allowed for in the *Bouguer anomalies* (so named after the French savant and explorer, Pierre Bouguer, 1749). If the attraction of the rocks between the site and sea level is called B, the calculated value of acceleration will be

$$g(\phi, h) + B$$

and the Bouguer anomaly will be

$$g_{obs} - g(\phi, h) - B$$

If, as is usual, the slope of the surface of the ground is small, the attraction of the rocks will be very close to that of an infinite slab of thickness h and density ρ, the density of the rocks. As is shown in example 2.3, the attraction of such a slab is

$$2\pi G \rho h$$

In mountainous country with steep slopes, it is necessary to calculate B more exactly from the actual form of the topography.

2.4 Local and regional variations

Bouguer anomalies are the residual variations after allowing for the known mass of rock above sea level and so ideally they should represent the gravitational attraction of rocks of varying density below sea level and should be explained by the geological structure if that is well enough known. In flat-lying country such as the British Isles the Bouguer anomalies can in fact be accounted for by the known geological structure, or they can be used to infer structure with some confidence. A very simple example is shown in figure 2.13. In the neighbourhood of Birmingham, coal-bearing rocks lie upon denser older harder rocks. If the thickness of the coal measures is t, the gravitational attraction is that of an infinite slab of thickness t and density $-\rho$, the difference between the density of the coal measures and that of the denser rocks. If the thickness varies only slowly from place to place, the attraction is

$$-2\pi G \rho t$$

and the Bouguer anomalies would be expected to be closely correlated with the thickness of coal measures. As figure 2.14 shows, that is in fact so.

Often the shapes of bodies of rock cannot be represented by infinite slabs and the attraction must be calculated either by taking a shape such as a cylinder as a model for which formulae are available that give the attraction, or by making a numerical calculation on a computer. Figure 2.15, for example, shows the observed and calculated attraction of a mass of granite which is less dense than the rocks that surround it.

The general problem of estimating the form of a structure that will produce a given variation of acceleration cannot be solved uniquely. Suppose that the density of matter in a given region is a function $\rho(r)$ of the position vector r measured from the point at which gravity is to be calculated. Suppose that the component of acceleration in the direction of a unit vector z is to be calculated. The attraction of matter in a volume V in the specified direction is

$$g = -\frac{\partial}{\partial z} \int \frac{G\rho(r)}{r} \, dV$$

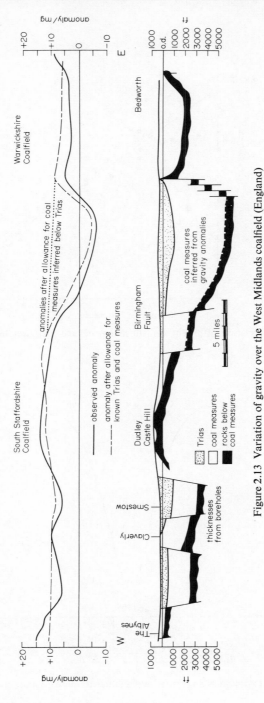

Figure 2.13 Variation of gravity over the West Midlands coalfield (England)

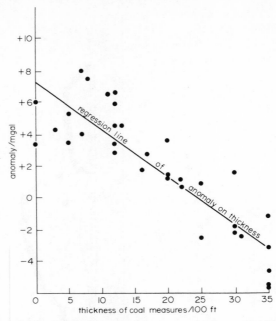

Figure 2.14 Correlation of gravity and thickness of coal measures in the West Midlands (England)

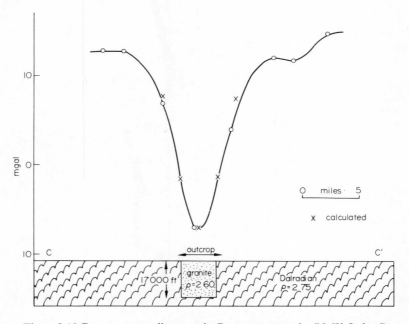

Figure 2.15 Bouguer anomalies over the Barnesmore granite (N. W. Ireland)

Now g is known as a function of position on the Earth and so this expression is an integral equation to be solved for the density as a function of position. In local surveys, the curvature of the Earth can be ignored and it may be supposed that the values of g are known over a horizontal plane. The integral equation would then be formulated in Cartesian coordinates. If an appreciable fraction of the surface of the Earth is being considered, acceleration must be considered as known over the spherical surface of the Earth and the integral equation will be formulated in spherical polar coordinates. The properties of the equation have been worked out in some detail by Bullard and Cooper (1948) and by Kreisel (1949) and they show that unique solutions cannot be found; in particular, if the observed values of g can be produced by a distribution of density at some depth below the surface, then they can also be produced by distributions of density at lesser depths. The converse is not true and there is a maximum depth at which the mass anomalies that will produce a given variation of acceleration may be placed. The reason is that, the more rapidly g varies, the greater the variation of density required at a specified depth and, below a certain depth determined by the maximum variation of g and the rate of variation with position, the necessary density variations become infinite. It follows that additional information beyond the values of g is needed to interpret gravity surveys in terms of geological structures. That information may be other geophysical information, such as the shapes of structures obtained from seismic surveys (chapter 3) or it may be geological information about likely values of density.

Formal solutions to the integral equation, in the form of density variations, can be obtained numerically, provided some suitable restriction, such as the depth of the structure, is imposed to enable an unique result to be obtained.

When the acceleration due to gravity is examined over large areas, and especially over mountainous regions, it is found that the simple behaviour found locally in areas of low relief is no longer followed. In fact, the actual variation is much better represented by the free-air anomalies than by the Bouguer anomalies. Curiously, this observation was first made by Bouguer (1749) himself in studying the behaviour of the plumb bob near Mt Chimborazo in the Andes and the general result was amply confirmed and widely studied by Archdeacon Pratt (1855) and Sir George Airy (1855) who both analysed observations made in the Himalayas, and by Hayford in the U.S.A. (1909). The early work was based on the *horizontal* attraction of a mountain as indicated by the deflection of a plumb bob; the results of measurements of the vertical component of the acceleration due to gravity are represented by those over the Alps shown in figure 2.16. The general result is that variations over mountains are less than ten per cent of the attraction of the mass of the mountains themselves, and the compelling inference is that the mass of the mountains above sea level is balanced by a deficit of mass at some depth. The general result is known as the *principle of isostasy*. The actual distribution of the mass deficiency will be discussed in the next chapter after the seismic evidence has been described.

A very exact isostatic balance also obtains between the oceans and the continents. On the face of it, the acceleration due to gravity over the oceans should be less than that over the continents by about 400 mgal corresponding to the difference between a column of water of density 1030 kg m^{-3} and thickness 5 km and an equal thickness of rock of density 2700 kg m^{-3}. Again, the actual variations as indicated in table 2.1 are less than ten per cent of that value and the interpretation of this result will also be considered after the seismic evidence has been set out.

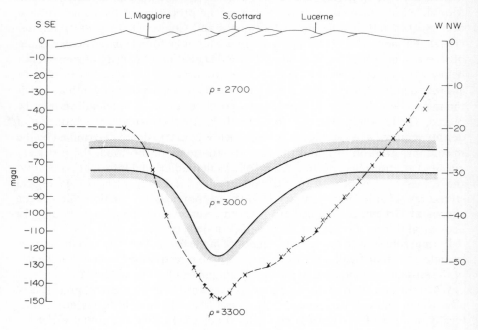

x— — x— —x Bouguer anomalies

Figure 2.16 Variation of gravity over the Alps. Right-hand scale: depth in kilometres below sea level

The study of variations of the acceleration due to gravity over the Earth as a whole is greatly hampered by the fact that so much of the surface is covered by seas. Despite the great advances in the technique of measurement of gravity at sea, the numbers and accuracy of data at sea are far less than those on land and

Table 2.1   Mean free-air anomalies over the oceans

	Anomaly (mgal)	
	Range	Mean without regard to sign
North Atlantic	+40 to −60	20
South Atlantic	+40 to −60	15
East Pacific		20
Mediterranean	+50 to −50	

so it is inevitable that world-wide averages are biased towards land values despite the fact that the area of the land is only a quarter of the whole surface. It is also difficult to separate the variations representative of the deep interior of the Earth when the measured values at the surface are strongly influenced by local shallow structures. Artificial satellites allow the restrictions of measurements at the surface to be avoided and provide unbiased estimates of the large-scale variations representative of the structure of the deep interior of the Earth.

2.5 Determination of the gravitational potential from the motions of artificial satellites

It has already been pointed out that the Earth is very nearly spherically symmetrical and that the gravity field around it is very nearly the same as that around a point mass. The orbit of a small object about such a mass point is well known to be an ellipse of fixed size and orientation in a fixed plane. The relation between the period, T, of the revolution of the satellite in its orbit and the length, a, of the semi-major axis of the orbit is given by

$$a^3 \left(\frac{2\pi}{T} \right)^2 = GM$$

where M is the mass of the Earth.

Commonly GM is written as μ, while $2\pi/T$, the mean angular velocity around the orbit, is denoted by n and called the *mean motion*.

GM is equal to 3.986×10^{14} m^3 s^{-2}.

Thus, if a is taken to be 6700 km, as for a satellite close to the Earth, T is 5460 s or 1/16 d (d stands for *days*).

Satellites used in television relays circle the Earth once in 24 h, so that they remain above the same point on the surface. For such *geostationary* satellites, $T = 1$ d = 86 400 s and so $a = 42$ 200 km.

It may be expected that artificial satellites will move around the actual Earth in paths which depart only slightly from fixed ellipses, and the equations of motion of artificial satellites should be formulated in a way that takes that into consideration.

Outside the Earth the gravitational potential satisfies Laplace's equation

$$\nabla^2 V = 0$$

which is nothing more than a statement of the inverse square law. Since the Earth is nearly spherically symmetrical, it is natural to express quantities in spherical polar coordinates, and Laplace's equation then reads

$$\frac{1}{r^2} \frac{\partial}{\partial r} \left(r^2 \frac{\partial V}{\partial r} \right) + \frac{1}{r^2 \sin\theta} \frac{\partial}{\partial \theta} \left(\sin\theta \frac{\partial V}{\partial \theta} \right) + \frac{1}{r^2 \sin^2\theta} \frac{\partial^2 V}{\partial \phi^2} = 0 \qquad (2.7)$$

where r is the radius measured from the centre of mass of the Earth, θ is the colatitude measured from the direction of the north pole, and ϕ is the longitude measured from the meridian of Greenwich.

It is shown in appendix 2 that solutions of Laplace's equation in spherical polar coordinates may be written as the product of functions of each of the co-ordinates separately:

$$V = R(r) . \Theta(\theta) . \Phi(\phi) \tag{2.8}$$

The three functions are

$$R(r) = r^{-n-1}$$

$$\Phi(\phi) = \cos m\phi, \sin m\phi$$

$$\Theta(\theta) = P_n(\cos\theta) \text{ (a Legendre function) for } m = 0$$

$$= P_n^m(\cos\theta) \text{ (an associated Legendre function) for } m \neq 0$$

The general solution to Laplace's equation may therefore be written as

$$-\frac{GM}{r} \left[1 - \sum_{n=2}^{\infty} \left(\frac{a}{r}\right)^n J_n P_n(\cos\theta) \right. $$
$$\left. - \sum_{n=2}^{\infty} \left(\frac{a}{r}\right)^n \sum_{m=1}^{m=n} (C_{nm} \cos m\phi + S_{nm} \sin m\phi) P_n^m(\cos\theta) \right] \tag{2.9}$$

or alternatively as

$$-\frac{GM}{r} \left[1 - \sum_{n=2}^{\infty} \left(\frac{a}{r}\right)^n J_n P_n(\cos\theta) - \sum_{n=2}^{\infty} \left(\frac{a}{r}\right)^n \sum_{m=1}^{m=n} J_{nm} P_n^m(\cos\theta) \cos(m\phi + \beta_{nm}) \right] \tag{2.9a}$$

M is the mass of the Earth and a is the equatorial radius.

Of the coefficients J_n, J_{nm}, J_2 is particularly important because it is connected with the moments of inertia of the Earth, a result known as McCullagh's theorem, which is derived in appendix 3. Because the Earth is almost symmetrical about the polar axis of rotation, all moments of inertia about axes in the equatorial plane (perpendicular to the polar axis) may be taken to be equal. Let them be denoted by A. The moment of inertia about the polar axis is distinct; let it be called C. Then McCullagh's theorem tells us that

$$J_2 = \frac{C-A}{Ma^2} \tag{2.10}$$

J_2 is closely related to the polar flattening and spin angular velocity of the Earth and is much the greatest of all coefficients in the general expression for the potential, for it is about 10^{-3}, whereas all the others are 10^{-6} or less.

Because the potential of the Earth departs from the simple potential of a mass point, the orientations of the planes in which the orbits of artificial satellites lie undergo small changes, as do the directions of the major axes and the eccentricities of the orbits. The calculation of these effects involves very heavy algebra because of the geometrical complexities of the orbit and its relation to an arbitrary harmonic term in the potential.

One of the simplest and, at the same time, one of the most important effects of the part of the potential corresponding to the equatorial bulge can be worked out by elementary means. Suppose that the orbit of the satellite is circular. Then, as Gauss showed, in order to find the changes in the orbit over many revolutions, the satellite may be replaced in imagination by a ring of the same radius spinning about the centre of the Earth. Such a ring would undergo gyroscopic precession which may be worked out from Euler's equations of motion for a rigid body (Landau and Lifshitz, 1969) (as shown in appendix 4). The geometry is shown in figure 2.17. E is the pole of the equator of the Earth and P is the pole of the orbit of the satellite. O is the centre of the Earth and EOP is the angle of inclination of the orbit to the equator (θ).

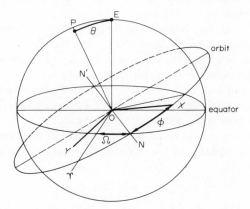

Figure 2.17 Geometry of a satellite orbit

The planes of the equator and the orbit intersect along the line of nodes, NN'. The direction of NN' is defined by the angle it makes to a fixed direction in the equatorial plane, the direction of a certain point in the constellation Aries, known as the first point of Aries (Υ). It is the direction in which the equator intersected the plane of the orbit of the Earth about the Sun at some past time. The angle between NN' and Υ is called the *longitude of the node* (Ω). If n is the angular velocity (or *mean motion*) of the satellite in its orbit, if r is the radius of the orbit and a the equatorial radius of the Earth, then

$$\dot{\Omega} = \frac{3}{2}n \, - \left(\frac{a}{r}\right)^2 J_2 \cos \theta \qquad (2.11)$$

The result shows that the plane of the orbit rotates slowly about the polar axis of the Earth in the direction opposite to the motion of the satellite. The orbits of all artificial satellites are dominated by this motion, for the attraction of the equatorial bulge is the largest departure from the central attraction of the total mass of the Earth.

With J_2 equal to 10^{-3} and n, for a close satellite, equal to 100 rad d^{-1}, the rate of motion of the node for a satellite at an inclination of 30° is 0·13 rad d^{-1} or 7·3° d^{-1}. Such a rate is readily detected even if a satellite is just observed with the naked eye against a background of stars.

More elaborate methods are needed to calculate the changes in the orbits of satellites produced by an arbitrary harmonic term in the potential. The general characteristics of the results will now be set out.

In the first place, harmonic terms in the series for the potential may be divided into three groups. There are two groups that do not depend on longitude— the *zonal* harmonics—and they are characterized by their symmetry about the equator; the *even* zonal harmonics are symmetrical and the *odd* zonal harmonics are antisymmetrical about the equator. The third group contains all the terms that depend on longitude.

The orbit of a satellite is described by a number of parameters called elements (figure 2.18). *The longitude of the node* (Ω) and the *inclination*, the angle θ between the plane of the orbit and the equator, have already been introduced. The angle measured in the plane of the orbit between the line of nodes and the direction of perigee, the position of closest approach of the satellite to the Earth, is called the *longitude of perigee* and is denoted by ω. The eccentricity of the orbit is denoted by e and the length of the semi-major axis by a_s.

The effect of the second zonal harmonic term in the potential is that the node moves along the equator in the opposite direction to the satellite at the rate of

$$-\frac{3}{2}n\left(\frac{a_e}{\eta^2 a}\right)^2 J_2 \cos\theta$$

(a_e is the equatorial radius of the Earth and $\eta^2 = 1 - e^2$).

All even zonal harmonics contribute to the steady motion of the node, and they all cause the longitude of perigee to change steadily, the rate for the second harmonic being

$$\frac{3}{4}n\left(\frac{a_e}{\eta^2 a_s}\right)^2 J_2(5\cos^2\theta - 1)$$

The odd zonal harmonics do not produce steady changes of any element but give rise to changes with a period equal to the rate at which perigee is caused to rotate by the even harmonics. The inclination and eccentricity undergo such oscillations as well as the node and perigee.

The harmonics which depend on longitude do not in general give steady or long-periodic changes in any element, but rather oscillatory changes in the elements with periods equal to $m(\tilde{\omega}t - \Omega')$, where Ω' is the longitude of the node measured from the observatory. If, however, the period of the satellite in its orbit is close to a simple rational fraction of the day, in which case the motion of the satellite is said to be *commensurable* with the rotation of the Earth, the elements of the orbit show changes with the speed

$$pn - q\tilde{\omega}$$

where p and q are small integers and $\tilde{\omega}$ is the spin angular velocity of the Earth. A *geostationary* orbit is one for which p and q are 1 and $n - \tilde{\omega}$ is very small; changes in the orbit caused by variations of the potential with longitude, in particular, any ellipticity of the equator, will have very long periods.

Steady and long-periodic variations of orbits are quite easy to determine with high accuracy because the observations can be continued over long intervals of time and because it is not necessary to know the position of the observatory from which they are made. The daily variations arising from the parts of the potential that depend on longitude are more difficult to determine because they

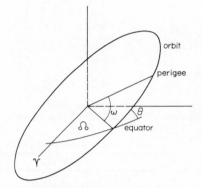

Figure 2.18 The elements of an orbit

also depend on the position of the observatory through Ω' and so cannot be obtained independently of improved positions for the observatory.

The steady rate of motion of the node of a satellite orbit is the sum of contributions from all even zonal harmonics and so may be written as

$$\dot{\Omega} = a_2 J_2 + a_4 J_4 + a_6 J_6 + \ldots + a_{2n} J_{2n} + \ldots \qquad (2.12)$$

where the a_{2n} are coefficients that depend on the elements of the orbit. Thus a_2 is

$$-\frac{3}{2} \frac{n}{\eta^4} \left(\frac{a_e}{a_s}\right)^2 \cos\theta$$

Although there are other contributions to the motion of the node, in particular, the attractions of the Sun and the Moon, they can be calculated with high accuracy, so that the rate of motion of the node on the left side of equation (2.12) is understood to be the observed rate corrected for the attractions of the Sun and the Moon and other known effects. There is one disturbance which, though large, has almost no effect on the motion of the node: the drag of the atmosphere on a close satellite causes the orbit to contract and to become more circular but causes only a very small rotation of the node.

The great difficulty in estimating the harmonic coefficients from observations of the motions of the nodes of orbits of artificial satellites is that there are more coefficients of comparable magnitude than there are distinct orbits. Orbits are distinguished by different values of eccentricity, semi-major axis and inclination but eccentricities are generally rather small and $1 - e^2$ is mostly very close to 1. Similarly, a_e/a_s is also close to 1, and indeed has to be if the effects of the higher-

Table 2.2   Coefficients of spherical harmonics in the expansion of the gravitational potential of the Earth

(a) Zonal harmonics			
$10^6 J_2$	1082·6	$10^6 J_3$	−2·54
J_4	−1·59	J_5	−0·21
J_6	0·50	J_7	−0·40
J_8	−0·12	J_9	0
J_{10}	−0·35	J_{11}	0
J_{12}	−0·04	J_{13}	0
J_{14}	−0·07	J_{15}	−0·20
J_{16}	0·19	J_{17}	0
J_{18}	−0·23	J_{19}	0
		J_{21}	0·26

(Values from Kozai (1969) and King Hele *et al.* (1969).)

(b) Tesseral and sectorial harmonics			
n	m	$10^6 C_{nm}$	$10^6 S_{nm}$
2	2	2·41	−1·36
3	1	2·0	0·26
	2	0·9	−0·6
	3	0·7	−1·4
4	1	−0·5	−0·5
	2	0·3	0·7
	3	−1·0	−0·2
	4	−0·08	0·3
5	1	−0·05	−0·1
	2	0·6	−0·4
	3	−0·4	−0·09
	4	−0·3	0·08
	5	−0·1	−0·6
6	1	−0·1	0·04
	2	0·05	−0·4
	3	0·03	0·04
	4	—	−0·4
	5	−0·2	−0·5
	6	−0·09	−0·07

(Some values from Gaposchkin and Lambeck (1970).)
Note. The coefficients of the spherical harmonics are 'normalized' so that the integral of the square of any surface harmonic over a unit sphere is 4π.

order harmonics are not to be negligible. Accordingly, significant variations of orbits only arise through differences of inclination, but the number of distinct inclinations is only about 10 because orbits of satellites launched from a particular site usually have very similar inclinations. Thus, despite the large number of satellites that have been launched altogether, only a limited number of coefficients can be estimated and those only by assuming somewhat arbitrarily that all beyond a certain order are zero. The lower-order harmonics may be considered to be well established, perhaps to $n = 6$ or 8 but, for the higher ones, only the order of magnitude is certain.

Similar considerations apply to the odd zonal harmonics, although it turns out that somewhat more are well known than the even harmonics. Again, all harmonics of a given degree m also contribute to variations in the orbital elements of speed $m(\tilde{\omega}t - \Omega')$ and so cannot be clearly separated one from the other.

Table 2.2 gives a list of some of the coefficients so far estimated. It is rather difficult to appreciate the significance of such a list and it is easier to understand the meaning of the results when they are presented as a map. The usual convention is to present a map of the mean shape of the sea-level surface, and the meaning of that idea must now be considered.

2.6 The shape of the Earth

The shape of the solid surface of the Earth is very irregular and of no general scientific interest. The shape of the Earth as understood in this section is the shape of the surface on which the sum of the potentials of the rotational acceleration of the Earth and of the Newtonian self-attraction is a constant, equal to the average value over the sea surface. The surface so defined is called the *geoid* and over the oceans it coincides with the mean sea-level surface, since the seas, if not subject to the tidal attraction of the Sun and the Moon nor to the actions of the winds, would flow until the surface was everywhere an equipotential of the total potential. There is some difficulty in defining the geoid over land areas, but to a first approximation it may be supposed to be the surface to which water in a canal cut through the land would flow. When we speak of the Earth as being a sphere flattened at the poles by about 1 part in 300, it is to the geoid that the description applies; when we speak of the Earth as having an equatorial radius of some 6378 km, it is again the radius of the geoid to which we refer; and when we say that the height of some place is so much above sea level, it is the height above the geoid that we mean. The geoid is the surface to which maps are referred (see Cook (1969) for a more detailed discussion of this point) and a good knowledge of the size and shape of the geoid is the basis of maps, while the shape itself, through its connection with the potential, provides information about the distribution of mass within the Earth.

The equatorial bulge dominates the form of the geoid, so that it is convenient to consider it separately from the smaller irregularities. Suppose that the geoid is an ellipsoid of revolution, symmetrical about the polar axis, and suppose that the radius vector measured from the centre at colatitude θ is r. Then

$$r = a(1 - f\cos^2\theta)$$

the polar equation of an ellipse referred to its centre. f is the polar flattening.

The potential of the Newtonian attraction is

$$-\frac{\mu}{r}\left\{1 - \left(\frac{a}{r}\right)^2 J_2 P_2(\cos\theta)\right\}$$

Since the seas are rotating with the Earth, any point in the sea surface is subject to the centrifugal acceleration

$$r\tilde{\omega}^2 \sin\theta$$

in the direction normal to the polar axis. The component along the radius vector is

$$r\tilde{\omega}^2 \sin^2\theta$$

which is the same as

$$-\frac{\partial}{\partial r}\left(-\tfrac{1}{2}r^2\,\tilde{\omega}^2 \sin^2\theta\right)$$

showing that the centrifugal acceleration may be regarded as derived from a potential

$$-\tfrac{1}{2}r^2\,\tilde{\omega}^2 \sin^2\theta$$

The total potential is thus

$$-\frac{\mu}{r}\left\{1 - \left(\frac{a}{r}\right)^2 J_2\,P_2(\cos\theta)\right\} - \tfrac{1}{2}r^2\,\tilde{\omega}^2 \sin^2\theta$$

it must be constant over the surface of the geoid. The condition for the potential to be constant is found by inserting the value of the radius vector on the geoid

$$r = a(1 - f\cos^2\theta)$$

$$= a\{1 + \tfrac{1}{3}f - \tfrac{2}{3}fP_2(\cos\theta)\}$$

It is convenient to put the ratio of the centrifugal to the gravitational acceleration at the equator equal to m, that is,

$$m = \frac{a^3\,\tilde{\omega}^2}{\mu}$$

m is about 3×10^{-3} or roughly $3J_2$.

Now in the expression for the potential it will be sufficient, in working to the first order of small quantities, to put a/r equal to 1 when it is multiplied by J_2

or m. The condition for the potential to be constant on the ellipsoid of revolution therefore reads

$$-\{1 - \tfrac{1}{3}f + \tfrac{2}{3}fP_2(\cos\theta)\} + J_2 P_2(\cos\theta) - \tfrac{1}{3}m\{1 - P_2(\cos\theta)\} = \text{constant} \quad (2.13)$$

using the relation

$$\sin^2\theta = \tfrac{2}{3}\{1 - P_2(\cos\theta)\}$$

The equation (2.13) can only be satisfied if the potential on the ellipsoid is independent of θ, that is if

$$-\tfrac{2}{3}f + J_2 + \tfrac{1}{3}m = 0 \qquad (2.14)$$

or

$$f = \tfrac{3}{2}J_2 + \tfrac{1}{2}m \qquad (2.14a)$$

The result is fundamental to the study of the shape of the Earth. As derived here, quantities of order f^2 have been neglected but a theory correct to second order is required in dealing with the actual Earth; the extension is straightforward.

With the numerical values

$$\tilde{\omega} = 7\cdot27 \times 10^{-5}\,\text{rad s}^{-1}$$

$$a = 6378\ \text{km}$$

$$\mu = 3\cdot986 \times 10^{14}\,\text{m}^3\,\text{s}^{-2}$$

m is found to be $3\cdot45 \times 10^{-3}$.

J_2 is known from satellite data to be $1\cdot08 \times 10^{-3}$; f is therefore $3\cdot35 \times 10^{-3}$.

The value of gravity over the ellipsoid of revolution is the standard with which actual values of gravity are compared in forming gravity anomalies. To the first order of such quantities such as f, g is $-\partial V/\partial r$. Hence if

$$V = -\frac{\mu}{r}\left\{1 - \left(\frac{a}{r}\right)^2 J_2 P_2(\cos\theta)\right\} - \tfrac{1}{2}r^2 \tilde{\omega}^2 \sin^2\theta$$

$$-\frac{\partial V}{\partial r} = -\frac{\mu}{r^2} + \frac{3\mu}{r^2}\left(\frac{a}{r}\right)^2 J_2 P_2(\cos\theta) + r\tilde{\omega}^2 \sin^2\theta$$

or

$$g = \frac{\mu}{r^2}\{1 - 3J_2 P_2(\cos\theta) - m\sin^2\theta\} \qquad (2.15)$$

As before, r is put equal to $a\{1 - \tfrac{1}{3}f + \tfrac{2}{3}fP_2(\cos\theta)\}$ in the leading term, and finally

$$g = \frac{\mu}{a^2}\{1 - \tfrac{1}{2}J_2 - \tfrac{5}{3}m + (2m - \tfrac{3}{2}J_2)\cos^2\theta\} \qquad (2.16)$$

which is of the form

$$g = g_e\{1 + \beta\sin^2\phi\}$$

where ϕ is the latitude and g_e is the value at the equator:

$$g_e = \frac{\mu}{a^2}\{1 - \tfrac{1}{2}J_2 - \tfrac{5}{3}m\} = \frac{\mu}{a^2}(1 - \tfrac{1}{3}f - \tfrac{3}{2}m) \qquad (2.17)$$

in addition

$$\beta = 2m - \tfrac{3}{2}J_2 = \tfrac{5}{2}m - f \qquad (2.18)$$

Again, the first-order theory is insufficient in practice, and one carried to the order of f^2 is needed. It is necessary to allow for the fact that ϕ, the geographical latitude, is not exactly $\tfrac{1}{2}\pi - \theta$, and that the acceleration due to gravity is perpendicular to the ellipsoid and is not directed along the radius vector:

$$g = \left\{\left(\frac{\partial V}{\partial r}\right)^2 + \frac{1}{r^2}\left(\frac{\partial V}{\partial \theta}\right)^2\right\}^{1/2}$$

The second-order theory gives a gravity formula

$$g = g_e(1 + \beta_1 \sin^2\phi + \beta_2 \sin^2 2\phi)$$

Given m and any one of the quantities J_2, f and β_1, the others may be found. The shape of the Earth can accordingly be determined from the external potential as found from artificial satellites, from measurements of gravity over the surface, or from survey measurements that give f directly; in the latter, the distance over the surface—the arc length along a meridian—is found from survey, while the direction of the normal to the geoid is found from the directions of stars relative to that of the plumb line. The shape is best determined from the external potential, mainly because the surface measurements are not uniformly distributed—the survey measurements cannot be made over the oceans, and gravity measurements at sea are fewer and less accurate than land gravity measurements. The values now adopted for f, β_1 and β_2 are therefore derived from J_2.

The size and mass of the Earth can of course be found from surface measurements but here again it is now better to use data from satellites. The product GM or μ can be found from the distance and period of a satellite and good results are available both from artificial satellites and from the Moon, the distance of which has been accurately measured by radar (see Yaplee *et al.*, 1965). The International Astronomical Union has adopted

$$a_e = 6378 \cdot 160 \text{ km}$$
$$\mu = GM = 3 \cdot 986 \times 10^{14} \text{ m}^3 \text{ s}^{-2}$$
$$J_2 = 1 \cdot 0827 \times 10^{-3}$$

and they entail

$$f = 1/298 \cdot 25$$
$$\beta_1 = 5 \cdot 3024 \times 10^{-3}$$
$$\beta_2 = -5 \cdot 9 \times 10^{-6}$$
$$g_e = 9 \cdot 780318 \text{ m s}^{-2}$$

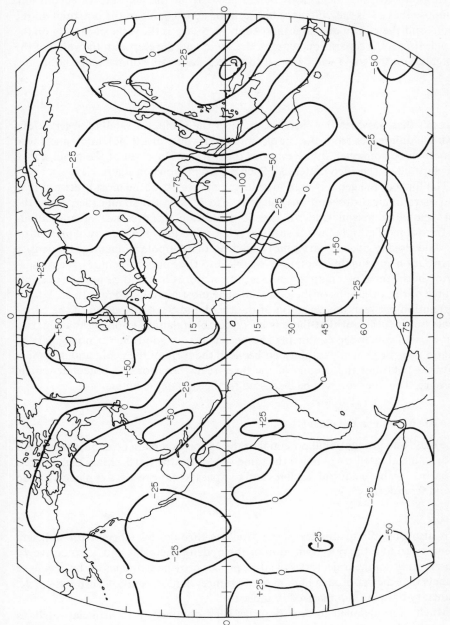

Figure 2.19 Map of the geoid, showing contours of equal departure, N (in metres), from a spheroid of flattening 1/298·25 (Gaposchkin and Lambeck, 1970)

The small terms in the potential represented by harmonics other than $P_2(\cos\theta)$ correspond to small deviations of the geoid from the ellipsoid of revolution. Suppose that δV is the difference between the actual value of the potential on the geoid and the ellipsoidal value, and suppose that g is the value of gravity on the geoid. If N is the separation between the geoid and the ellipsoid, the value of the potential on the geoid will be equal to the constant value on the ellipsoid if

$$N = - \frac{\delta V}{g}$$

a result that provides the basis for maps of the geoid, for satellite observations give the differences from the simple ellipsoidal value of the potential, and those may then be converted into elevations (positive or negative) of the geoid above an ellipsoid of specified flattening. Such a map is shown in figure 2.19.

The form of the geoid is used in reducing the results of the most precise survey and astronomical observations but it is also of physical significance, since the corresponding irregularities of potential indicate the presence of anomalous distributions of mass below the surface. While this is just another way of presenting the data already given as harmonic coefficients or multipole moments of the density distribution and the interpretation is subject to the same ambiguities, the presentation in the form of a map sometimes gives a more direct indication of a possible source of the variations of the potential. The most obvious thing is to look for a correlation between the elevation of the geoid and the distribution of the continents but it turns out that there is no detectable relation, implying that the isostatic balance between continents and oceans is very good. So far no significant relation has been firmly established between the form of the geoid and any other physical field and the origins of the large-scale variations of potential remain obscure. The question will be discussed further in later chapters.

2.7 The Moon and the planets

The mass and gravitational potential of the Moon or a planet can be investigated with artificial satellites in much the same way as for the Earth. Mars and Jupiter, in particular, have natural satellites and the masses of the respective planets can be found from Kepler's law

$$\mu = n^2 a^3$$

It is also possible to observe some satellites through telescopes with sufficient accuracy to find the rate of motion of the node of the orbit and so to derive J_2. Observations of the trajectories of space probes provide estimates of the masses of planets that they approach and, in one instance, that of Venus, an estimate of J_2. Planetary data are summarized in table 11.1.

Much more detailed information has been obtained from artificial satellites about the Moon. Their orbits undergo changes related to the potential of the Moon so that harmonic coefficients may be determined, as indicated in table 2.3, but in addition it has been possible to make direct observations of the acceleration due to gravity above part of the surface. Changes in velocity of a satellite along the line of sight to the Earth give rise to changes (Doppler shifts) in the frequencies of radio signals emitted by the satellite and received at the Earth. The change in the received frequency over a given time is a measure of the acceleration of the satellite along the line of sight, so that the variation of acceleration of a satellite close to the surface of the Moon can be derived. In this way it has been found that there are considerable local concentrations of mass near the surface (*mascons*) associated with certain of the lunar maria. Harmonic analysis based on long-periodic changes of the orbits of satellites gives much the same results as the direct Doppler observations, but is in some ways more informative because the Doppler results are confined to the visible side of the Moon and are most reliable near the centre, whereas data derived from changes in the orbits apply to the whole surface.

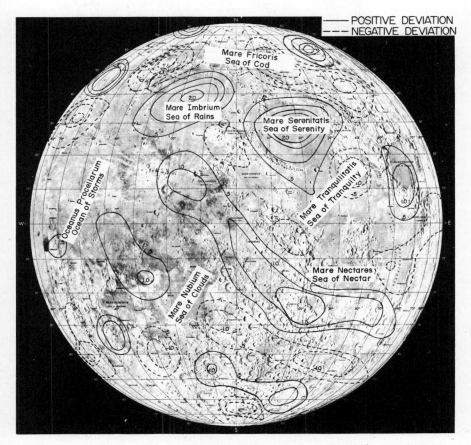

Figure 2.20 Map of gravity at the surface of the Moon: near side (Michael *et al.*, 1969)

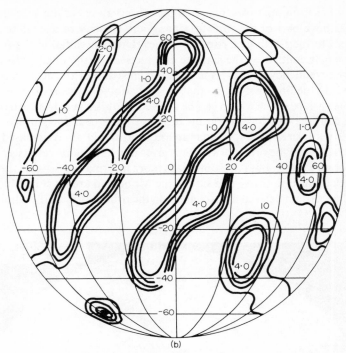

(b)

Figure 2.21 Map of gravity at the surface of the Moon: far side (Michael *et al.*, 1969)

Table 2.3   Coefficients of spherical harmonics in the expansion of the gravitational potential of the Moon

n	m	$10^4 C_{nm}$	$10^4 S_{nm}$
2	0	−0·96	—
	1	−0·06	−0·01
	2	0·34	0·20
3	0	−0·08	—
	1	0·34	0·07
	2	−0·08	−0·06
	3	−0·19	−0·36
4	0	0·03	—
	1	−0·13	0·06
	2	0·16	0·02
	3	0·27	0·46
	4	0·43	0·37
5	0	−0·05	—
6	0	−0·03	—
7	0	+0·04	—
8	0	−0·05	—

(All coefficients are normalized—see the note to table 2.2.)

Maps showing the form of the lunar potential are given in figures 2.20 and 2.21 while table 11.1 contains values of J_2 for the Moon and planets. The interpretation of these data will be discussed in chapter 5.

Examples for chapter 2

2.1 In a determination of the acceleration due to gravity by the N.P.L. method (figure 2.2) the interval between the slits was 1·000 05 m and the time intervals between the upward and downward crossings were
   at the upper level, 0·204 05 s
   at the lower level, 0·925 75 s
   Calculate the acceleration due to gravity.

2.2 The period of an invariable pendulum was 0·507 201 s at a site where the acceleration due to gravity was 9·811 63 m s^{-2} and 0·507 120 s at a second site. Calculate the acceleration due to gravity at the second site.

2.3 Show that the gravitational attraction of an infinite slab of thickness t and density ρ is $2\pi G \rho t$.
   (Show that the vertical attraction exerted by a vertical cylindrical shell of height t, radius r and thickness dr at a point h above the upper surface is

$$2\pi G\rho th \, \frac{r \, dr}{(r^2 + h^2)^{3/2}}$$

Then, by integrating from 0 to r, show that the attraction of the slab is $2\pi G\rho t$.)

2.4 Coal measures 1800 m thick and of density 2510 kg m^{-3} are surrounded by older rocks of density 2750 kg m^{-3}. By how much is the acceleration due to gravity less over the middle of the coal measures than over the surrounding older rocks? ($G = 6\cdot67 \times 10^{-11}$ N m^2 kg^{-2}.)

2.5 A ship steams westwards at 25 km h^{-1} in latitude 30°. By how much does the acceleration measured on the ship differ from the value measured if the ship were stationary? The radius of the Earth is approximately 6400 km.

2.6 The harmonic terms proportional to $P_3(\cos\theta)$ and $P_{15}(\cos\theta)$ have equal coefficients, and therefore equal amplitudes at the surface of a planet. Compare their amplitudes at the heights of 500 and 1500 km above the surface. Take the radius of the Earth to be 6400 km.

2.7 What is the period of a satellite in an orbit of semi-major axis 8000 km:
   (a) around the Earth ($M = 6 \times 10^{24}$ kg):   120 m (2 h)

   (b) around the Moon $\left(\dfrac{\text{mass of Moon}}{\text{mass of Earth}} = \dfrac{1}{81}\right)$: 1080 m (18 h)

2.8 Calculate J_2 for Mars from the following data:

satellite Phobos period	0·32 d
semi-major axis	9000 km
inclination	1°·1
regression of node	0°·56 d^{-1}
equatorial radius of Mars	3375 km

Further reading for chapter 2

Bomford, G., 1970, *Geodesy*, 4th edn (Oxford: Clarendon Press), viii + 731 pp.

Bullard, E. C., and Cooper, R. I. B., 1948, 'The determination of the masses necessary to produce a given gravitational field', *Proc. Roy. Soc.*, Ser. *A*, **194**, 332–47.

Cook, A. H., 1967, *Phil. Trans. Roy Soc.*, Ser. *A*, **261**, 211–52.

—— 1969, *Gravity and the Earth* (London: Wykeman Press), xi + 95 pp.

Dollfus, A. (Ed.), 1970, *Surfaces and Interiors of Planets and Satellites* (London, New York: Academic Press), xi + 569 pp.

Gaposchkin, E. M. and Lambeck, K., 1970, *1969 Smithsonian Standard Earth (II)*, *Res. Space. Sci. Smithsonian Astrophys. Obs. Spec. Rept.*, No. 315, May.

Hammond, J. A., and Faller, J. E., 1972, 'Precision measurements and fundamental constants', *Natl. Bur. Stand. Spec. Publ.*, No. 343, 457–63.

King Hele, D. G., Cook, G. E., and Scott, Diana W., 1969, 'Evaluation of odd zonal harmonics in the geopotential, of degree less than 33, from the analysis of 23 satellite orbits', *Planetary Space Sci.*, **17**, 629–64.

Kozai, Y., 1969, 'Revised values for coefficients of zonal spherical harmonics in geopotential', *Smithsonian Astrophys. Obs. Spec. Rept.*, No. 295.

Kreisel, G., 1949, 'Some remarks on integral equations...', *Proc. Roy Soc.*, Ser. *A*, **197**, 160–83.

Landau, L. D., and Lifshitz, E. M., 1969, *Course of Theoretical Physics—I, Mechanics* (Oxford, London: Pergamon Press), vii + 165 pp.

Michael, W. H., Blackshear, W. T., and Gapcynski, J. P., 1969, 'Results of the mass and gravitational field of the Moon as determined from the dynamics of lunar satellites', XII *COSPAR Conf.*, Prague.

Sakuma, A., 1972, 'Precision measurements and fundamental constants', *Natl. Bur. Stand. Spec. Publ.*, No. 343, 447–56.

Tucker, R. H., Cook, A. H., Iyer, H. M., and Stacey, F. D., 1970, *Global Geophysics* (London: English Universities Press), 199 pp.

Worzel, J. L., and Harrison, J. C., 1966, 'Gravity at sea', in *The Sea*, Vol. III (Ed. M. N. Hill) (London: Interscience), xvi + 963 pp.

Yaplee, B. S., Knowles, S. H., Shapiro, A., Craig, K. J., and Brouwer, D., 1965, 'The mean distance to the Moon as determined by radar', in *I.A.U. Symposium No. 21—The System of Astronomical Constants* (Ed. J. Kovalevsky) (Paris: Gauthier-Villars), 330 pp.

CHAPTER THREE

Elastic waves in the Earth

'Then the Earth shook and trembled,
The foundations also of the mountains moved.'
Psalm 8

3.1 Effects of earthquakes and large explosions

An earthquake occurs when forces build up within rocks to such an extent that
they exceed the strength of the rock and a break takes place, usually by a crack
appearing across which rocks suddenly slide over each other. The shock is felt at
greater or smaller distances according to the strength of the earthquake and, as
is well known, the larger earthquakes can cause movements of a metre or more
at the surface at places close to the site of the earthquake. The damage caused
by an earthquake usually arises from the effects of the accelerations of the ground
upon buildings or from the disturbance of soft unconsolidated and unstable
sand or clay. An earthquake in the sea bed or near the sea shore generates a wave
in the water which may produce a large surge when it enters shallow water with
very damaging consequences; such waves are called *tsunamis* and a special service
exists to give warning of their approach in the Pacific.

Earthquakes occur for the most part close to the surface because there the
rocks are strongest and the largest stresses can build up before a break occurs.
At greater depths, where the rocks are hotter, they can in general withstand
less stress and yield by flowing like treacle instead of breaking; however, in
certain regions of the Earth, earthquakes, known as deep focus earthquakes,
occur at depths as great as 600 km. Figures 3.1 and 3.2 are maps of the distribution
of shallow and deep earthquakes over the world, and it will be seen that they
occur almost entirely in well-defined belts, the significance of which will be
discussed in chapter 10.

Large explosions, especially of underground nuclear bombs, differ from
earthquakes in that they produce a uniform pressure rather than a shearing force
and so their effects at the surface are somewhat different from those of earthquakes. Of course, in testing nuclear bombs underground, care is taken to bury
the explosive deep enough that no serious effects occur at the surface but small
disturbances do show themselves, particularly as further breaks along lines of
existing geological weakness.

The local effects of earthquakes can be measured in a rough way by the type of
damage that they do—for example, tiles may be shed from roofs or pictures be
disturbed on the walls of houses, and, by mapping out places where similar

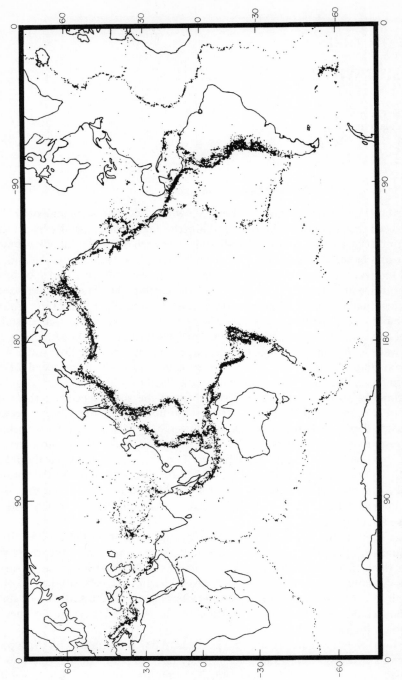

Figure 3.1 World-wide distribution of shallow earthquakes

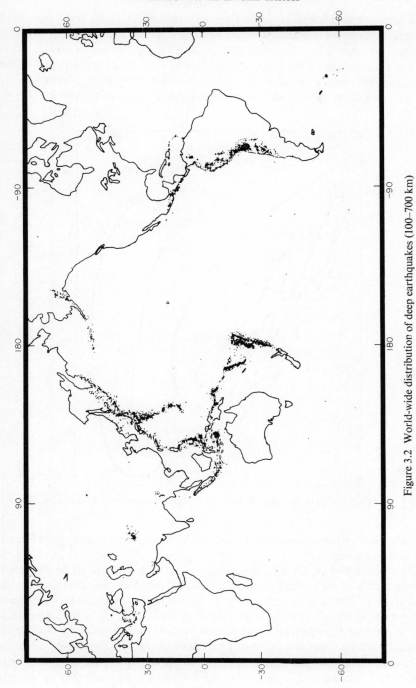

Figure 3.2 World-wide distribution of deep earthquakes (100–700 km)

effects occur, it is possible to draw lines of equal effect. Such lines are called isoseismal lines and are usually ellipses centred over the site of the earthquake, the *epicentre* (figure 3.3). The surface effects of an earthquake are by no means the same for earthquakes of the same strength for they depend on the depth of the earthquake, the types of rock at the surface, whether they already have lines of weakness, and so on. Nonetheless, it is useful to have a scale of the surface effects,

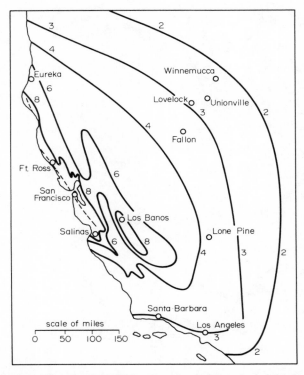

Figure 3.3 Isoseismal lines for the 1906 San Francisco earthquake (see Tucker *et al.*, 1970)

especially for drawing isoseismal lines, and such a scale is provided by the Mercalli scale, given in appendix 5.

If more quantitative ideas of the effects of earthquakes are to be obtained, the movements of the surface must be measured. Buildings are damaged by the acceleration of the ground, especially by horizontal accelerations which give rise to shearing displacements of the parts of a building. It is therefore most useful to measure the acceleration of the ground and that can be done close to an earthquake with a relatively insensitive accelerometer. As the effects of earthquakes are followed to greater and greater distances, more sensitive instruments are of course needed, and it may be better to measure displacement or velocity rather than acceleration. The designs of the appropriate instruments are discussed in the next section.

When the effects of an earthquake or explosion are observed at increasing distances, it is found that the time to the initial movement of the ground is proportional to the distance from the earthquake, at least when the earthquake is close enough that the curvature of the Earth may be neglected. The first impulse may last only for a second or so, and is then followed by a second impulse, again after a time proportional to the distance from the earthquake. Study of the ground movements in the two impulses shows that the first, or *primary*, pulse, denoted by *P*, corresponds to a wave motion with movement in the direction of propagation, while the second corresponds to a wave with motion at right angles to the direction of propagation. The second pulse is called the *secondary*, or *S*, phase. As will be seen in more detail below, further pulses, transmitted by a variety of paths, are seen at greater distances. The theory of the wave motion involved is given in section 3.3.

Waves from small chemical explosions are used commercially to investigate geological structures that may be of economic importance, while small earthquakes and the larger chemical explosions may be used to study the crust and upper mantle. Waves from the largest earthquakes and nuclear explosions travel right through the Earth and probe its deepest structure.

3.2 Seismometers

Suppose that a mass is suspended from an elastic framework fixed to the Earth; if the ground moves, then the mass, on account of its inertia, will at first remain at rest relative to the centre of the Earth, and therefore the relative position of the mass and the ground will change. The framework will accordingly be distorted and will be set into vibrations. The differences between seismometers arise from the nature of the suspension and especially whether it is designed to enable vertical or horizontal movement to be measured; from the system used to measure the relative displacement of the ground and the mass; and from the property of the motion—displacement, velocity or acceleration—that it is desired to measure. Seismometers may also be placed on a fixed site in an observatory or they may be required for use in temporary sites in the field for exploration work—the two types of instrument need different designs.

Figure 3.4 shows the principle of a seismometer for the measurement of horizontal movement. A mass is supported at the end of an arm that is pivoted about a nearly vertical hinge.

Then, as shown in appendix 6, the arm swings about the pivot with an angular frequency ω given by

$$\omega^2 = \frac{gr \sin \phi}{r^2 + k^2}$$

where r is the distance of the centre of mass from the hinge, k is the radius of gyration of the mass, g is the acceleration due to gravity and ϕ is the inclination of the hinge to the vertical. By making ϕ very small, the period $2\pi/\omega$ may be made much greater than $2\pi(r/g)^{1/2}$, the period of a simple pendulum of the same length, r.

The general solution for the response to a periodic displacement of the form

$$u = u_0 e^{ipt}$$

$\theta = \theta_0 e^{ipt}$ with

$$\theta_0 = \frac{\alpha u_0 p^2}{\omega^2 - p^2}$$

Here α is $r/(r^2 + k^2)$.

There are two extreme cases. In the first, $\omega \ll p$ (long-period instrument),

$$\theta_0 = -\alpha u_0$$

and the angular movement of the beam is a measure of the ground displacement, u.

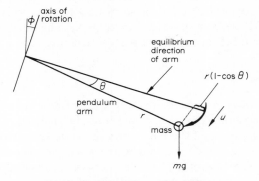

Figure 3.4 The horizontal pendulum

In the second case, $\omega \gg p$ (short-period instrument),

$$\theta_0 = \alpha u_0 p^2 / \omega^2$$

and the movement of the beam is a measure of the acceleration of the ground, $-p^2 u$.

A system in which the rotation of the beam is measured directly can therefore be arranged to measure either the displacement or the acceleration of the ground. A suitable electrical system for recording the displacement of the beam might consist of a differential transformer, in which a coil attached to the beam moves between two fixed coils. If the latter are excited with alternating voltages in opposite phases, the voltage induced in the beam coil will be approximately proportional to the displacement of the arm from its position of equilibrium and will change sign as the arm passes through that point. The induced voltage may be recorded electrically as a function of time.

Depending on the period of the motion it is desired to record, a seismometer will be arranged to record the displacement or the velocity or the acceleration of the ground.

The ground is continually in movement as a result of both human and natural forces. The former are due to machines, trains, cars and so on; the periods are about 0·1 s, or shorter than most natural seismic signals and so the accelerations are relatively large compared with those of the desired signals. It would then be desirable to record displacements. However, it is usually possible to avoid man-made noise by choosing the site of the observatory properly, and the main limitation to the detection of a seismic signal is the effect of natural noise, or microseisms. Most such noise is generated by storms at sea and is most intense near the coasts. In the west of Europe the amplitudes are of the order of 1 μm and the periods 5 to 10 s, periods that are long compared with the short-period seismic signals; an accelerometer would then be used, but a displacement meter would be best for seismic signals with periods longer than the microseismic periods.

Over a wide range of periods, an instrument that measures the velocity of the ground motion is convenient. Suppose that the fixed coils in the pick-up system just described are fed with a steady current so that they produce a constant magnetic field at the coil on the arm. The voltage induced in that coil will then be proportional to $\dot{\theta}$, and if $\omega \ll p$ it will be proportional to $\dot{u}$.

Electrical recording systems thus enable the simple horizontal pendulum to give a response proportional to the displacement, velocity or acceleration of the ground motion of dominant period, and the response may be further modified if damping is applied, either mechanically or with electrical feedback. Displacement meters suffer from one disadvantage that sometimes leads to velocity meters being adopted instead. If ω is to be small (long period), then ϕ also should be small, as indeed it should also be for high acceleration sensitivity. Mechanically, a pendulum with the hinge very nearly vertical is rather unstable and the position of equilibrium changes slowly. That is not important for velocity or acceleration meters but with a displacement meter it means that the response may exceed the range of the recorder. The difficulty can often be overcome by applying electrical feedback with a time constant long compared with that of the signal it is desired to record.

The spring gravity meter with the zero-length spring described in chapter 2 also behaves as a seismometer. Very similar considerations apply as to the horizontal pendulum and the instrument may be made to record vertical acceleration, velocity or displacement, and feedback may be applied to damp the system or to stabilize the equilibrium position of a displacement meter.

It has been supposed in the above description that the electrical recording system does not react back upon the mechanical system, and modern electronic systems are so sensitive that that can be a very good assumption. It was not true for the first electrical recording systems in which a coil attached to suspended mass moved in a permanent magnetic field and was connected to the coil of a moving-coil galvanometer. The equation of motion of the two coupled oscillatory systems are complex, but the parameters of the systems can be adjusted to give a wide range of responses to the ground motion. In particular, a mechanical system of short period coupled to a galvanometer of long period gives a good response to ground motions with long periods.

The seismometers so far considered are suitable for installations in fixed observatories but are not practical for use in the field. For that purpose, stiffer

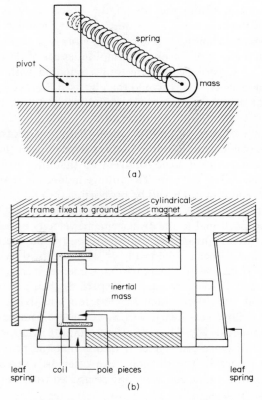

Figure 3.5 Some designs of seismometer: (a) long-period vertical; (b) short-period horizontal

and more compact systems are needed so that they may easily be carried from place to place without suffering damage. Some common designs are indicated in figure 3.5.

It is difficult to make seismometers of the foregoing types with periods greater than a few tens of seconds, yet there are seismic signals of very much longer period. Such signals can be detected with instruments that measure the strain of the ground (that is the change of displacement over some distance) rather than the displacement at a point. The principle of a strain meter is indicated in figure 3.6.

A horizontal rod of fused silica (mechanically stable and with a low coefficient of thermal expansion) is rigidly attached to the ground at one end, while the other end is suspended freely. The displacement of the free end relative to the ground at that point is measured electrically, for example, with a differential capacitor, and is equal to the strain multiplied by the length of the rod. With a rod 100 m long, the displacement of the free end may be some micrometres; the instrument is commonly used to detect signals having periods of hundreds of seconds.

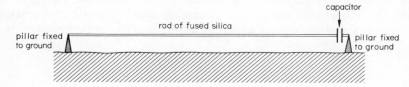

Figure 3.6 Benioff strain gauge using a rod of fused silica

The fused silica strain gauge, in common with all other seismometers, is sensitive to temperature and to atmospheric pressure, both of which change the overall length of the rod. Recently, a strain gauge that is in principle absolute and of high sensitivity has been developed. The light emitted by a helium–neon gas laser has such a narrow spectrum that sharp interference fringes can be produced in a Michelson interferometer having a path difference of many hundreds of metres. Figure 3.7 shows how such an interferometer may be set up to measure the change of strain of the ground. The changes in the length of the long path, reflected in the number of interference fringes passing across the detector, are recorded electrically.

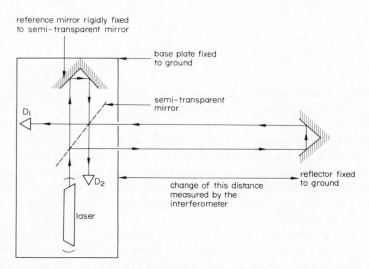

Figure 3.7 Laser interferometer strain gauge: D_1, D_2, photodetectors

Much attention has been paid to the design of seismometers in recent years, especially for the detection of nuclear explosions, and instruments are available to measure ground motion with periods covering the range from 0·01 s to hundreds of minutes and with amplitudes ranging from 0·1 μm to 1 mm. Nonetheless, microseisms set a severe limit on the detection of individual events but the relative noise levels can be greatly reduced by combining the signals from a number of seismometers in a coherent way. The principle is shown in figure 3.8. Seismometers are set out in an array consisting of two lines more or less at right angles. The instruments are at equal distances apart. Suppose that a seismic signal passes across the ground so that its speed along one of the lines is v. If the

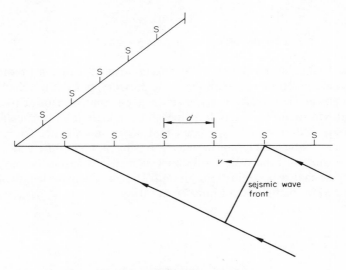

Figure 3.8 Seismometer array: S, seismometers

separation of the seismometer is d, the signals at the successive seismometer will be delayed by times v/d so that, if they are combined with such delays, the signal will be enhanced at the expense of the microseisms which in general will show no such consistent behaviour over the array. v depends on the type of seismic signal and the direction of the source, and is not known for any particular signal but, if the records from the individual seismometers are stored in a computer, they may be combined with different delays until the best result is found. Results from the two lines give both the speed of the signal over the ground and the direction from which it arrives.

While a single seismometer gives times of arrival of signals, the character of the ground motion can only be inferred if three seismometers, sensitive to the vertical and the two horizontal components of the motion, are set up together; the combination will also give improved discrimination against noise.

3.3 Elastic waves

When a shock is given to some point of an elastic solid, waves of disturbance spread out from the point, some of them moving through the body of the solid (bodily waves) and others being constrained to follow any surface of discontinuity, in particular a free surface. Most of the information about the interior of the Earth obtained from seismic observations has come from the bodily waves, the theory of which is given in appendix 7. The theory of surface, or guided waves, is given in section 3.6 and appendix 10.

The results of appendix 7 show that two types of elastic wave can be propagated through an isotropic solid, provided the material behaves linearly, the strains being small and the stresses well below the elastic limit of the material. The speed of the dilatational wave, usually denoted by α, is the greater, being equal to $\{(K + \frac{4}{3}\mu)/Gk\rho\}^{1/2}$, where K is the bulk modulus, μ the shear modulus and ρ the density. The speed of the shear waves, β, is $(\mu/\rho)^{1/2}$. Pulses of dilatational waves arrive first at a detector and are known as the P phases, P standing for primary, although many seismologists think of them as 'push–pull'. Pulses of shear waves, which arrive later, are known as the secondary, or S phases, often thought of as 'shake' waves.

A liquid has no shear strength and the shear modulus is zero. Shear waves are not transmitted by a liquid, but only dilatational waves.

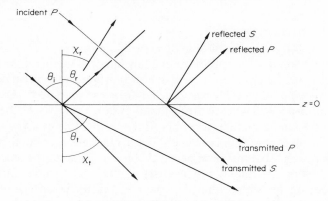

Figure 3.9 Reflection and refraction of elastic waves

Elastic waves, like light or electrical waves, are reflected and refracted at the boundary between different media. Let two media be in contact along the plane $z = 0$ (figure 3.9) and consider a dilatational wave incident from the upper medium (1) and making an angle θ_i with the normal to the surface. At the common surface, the displacements must be the same on the two sides and the stresses must be the same. Then, as shown in appendix 8, reflection and refraction take place according to a generalized form of Snell's law:

$$\frac{\sin \theta_i}{\alpha_1} = -\frac{\sin \theta_r}{\alpha_1} = -\frac{\sin \chi_r}{\beta_1} = \frac{\sin \theta_t}{\alpha_2} = \frac{\sin \chi_t}{\beta_2}$$

where α_1, α_2, β_1 and β_2 are the dilatational and shear velocities in the two media; θ_r and θ_t are the angles of reflection and refraction of the dilatational waves; χ_r and χ_t are the angles of reflection and refraction of the shear waves.

In the general case, none of the amplitudes is zero, so that two types of reflected and two types of refracted wave are generated from an incident dilatational wave. Similar results are found for an incident shear wave. If the second medium is a liquid no shear wave is propagated.

An important case is that when θ_t is $\pi/2$ so that the transmitted dilatational wave travels parallel to the boundary between the two media. Then $\sin \theta_i = -\sin \theta_r = \alpha_1/\alpha_2$ and θ_i and θ_r are real angles less than $\pi/2$ if $\alpha_1 < \alpha_2$. Since β_2 is less than α_2, χ_t is also a real angle less than $\pi/2$.

Far from a source, diffraction effects may be ignored and the paths of waves may, as for light or sound waves, be worked out geometrically by considering rays perpendicular to the wave fronts which, in a uniform medium, will be plane. These conditions generally apply at the surface of the Earth within distances from the surface small compared with the radius of curvature of the Earth. Over greater distances, account has to be taken of the fact that the elastic properties of the Earth vary with radial distance from the centre, so that surfaces of constant velocity are spheres centred on the centre of the Earth.

3.4 The outer layers of the Earth

When the signals due to an explosion of a few kilogrammes of TNT are observed with a seismometer at distances of up to a few hundred metres, it will be found that the first signal occurs after a time that is proportional to the distance from the explosion, showing that a wave has travelled from the source to the seismometer at a constant speed. The wave to arrive first is always a P wave. As the distance is increased it is found that, after a certain distance, the time of arrival of the first pulse is no longer proportional to the distance from the source but to a somewhat greater distance and that the speed of travel is greater than for the pulse at shorter distances. Figure 3.10 shows a typical set of data. Such observations can be accounted for on the supposition that the rocks near the surface of the Earth are arranged in horizontal layers with seismic velocities increasing with depth. When waves travel in a deeper layer, they do so at a higher speed than in the overlying layer but, because they have to travel down to and up again from the lower layer, they are delayed as compared with those in the upper layer but, beyond a certain distance, the greater speed compensates for the delay. The model of a set of layers of distinct properties is a good one for the outer parts of the Earth provided the depth, or thickness, of a layer is very small compared with the radius of the Earth so that the effects of curvature may be ignored and changes of properties with pressure are unimportant. In practice, the Earth may be investigated on this basis to a depth of nearly 100 km.

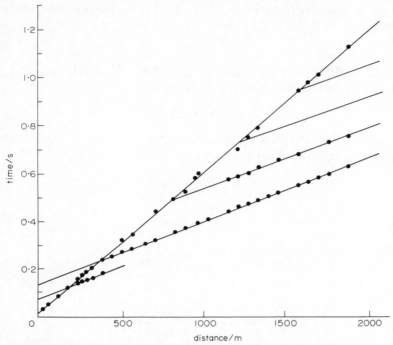

Figure 3.10 Example of observed travel–time data from a relatively close explosion

Figure 3.11 shows a set of layers of rocks with successively greater seismic velocities at increasing depths. The layers are numbered from the top down. Let v_i be the velocity in layer i, the thickness of which is h_i. Let D be the distance between the source and the detector at the surface. Consider the wave that travels down from the source at such an angle θ to the vertical that on refraction

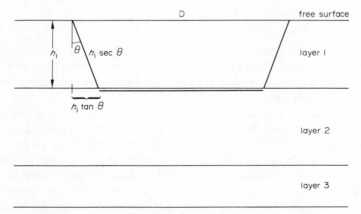

Figure 3.11 Geometry of seismic rays in two-layer model

at the boundary between layers 1 and 2, it travels along the boundary. Then by Snell's law

$$\sin \theta = v_1/v_2$$

Now the distance from the source to layer 2 is $h_1 \sec \theta$, while the distance travelled along the boundary between the layers is

$$D - 2h_1 \tan \theta$$

The time taken by the wave to travel down to layer 2 through layer 1, along the boundary in layer 2 and up again to the surface in layer 1 is then

$$\frac{2h_1 \sec \theta}{v_1} + \frac{D - 2h_1 \tan \theta}{v_2}$$

Since

$$\sec \theta = \frac{v_2}{(v_2^2 - v_1^2)^{1/2}} \quad \text{and} \quad \tan \theta = \frac{v_1}{(v_2^2 - v_1^2)^{1/2}}$$

the time is

$$\frac{D}{v_2} + 2h_1 \left\{ \frac{v_2}{v_1} \frac{1}{(v_2^2 - v_1^2)^{1/2}} - \frac{v_1}{v_2} \frac{1}{(v_2^2 - v_1^2)^{1/2}} \right\}$$

or

$$\frac{D}{v_2} + 2h_1 \left(\frac{1}{v_1^2} - \frac{1}{v_2^2} \right)^{1/2}$$

This result shows that, if the times are plotted against distance, they lie on a straight line with a slope of $1/v_2$ making an intercept of

$$2h_1(v_1^{-2} - v_2^{-2})^{1/2}$$

upon the axis of time. Thus, if v_1 is known (from observations at shorter distances where the waves travel only in the upper layer), v_2 and h_1 may be found.

The formula is easily extended to more than two layers arranged as shown in figure 3.12 (example 3.6).

While the model of a set of horizontal layers of rock gives a good first approximation in many actual circumstances, there are also important departures which may make it difficult to calculate thicknesses and velocities from the formula just given. In general the layers may slope and then observations made with the seismometers always up the slope or down the slope from the source will give erroneous results; observations should always be made if possible with seismometers on both sides of the sources. Layers are not commonly uniform and variations of thickness or velocity, particularly locally under a source or detector, cause the measured times to depart from the best fitting straight line. Lastly, the plot of times against distance is usually far more complex than the foregoing theory would suggest. According to the simple model, the first arrival should be followed by later signals that have travelled through the successively shallower and slower layers but, in practice, many more later arrivals are often seen. Evidently the interpretation of seismic observations needs some care and, in particular, statistical tests should be applied to assess the significance of any interpretation.

The model of a set of layers of rocks does, however, seem to represent the real disposition of rocks in very many cases, and works well when the boundaries are indeed nearly horizontal, although when there are appreciable slopes the interpretation may be ambiguous. The early work of Bullard *et al.* (1940) in East Anglia is a good example of the study of geological structure in favourable circumstances, the different rocks being in general quite distinct in their velocities

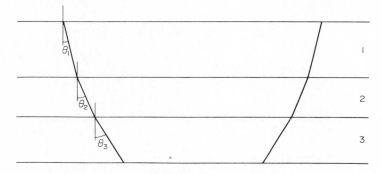

Figure 3.12 Geometry of seismic rays in multiple-layer model

and lying in more or less horizontal layers. An example of one of their records is shown in figure 3.13 and a section of the structure of East Anglia is given in figure 3.14. As figure 3.13 shows, the pulse refracted from the Palaeozoic floor is usually clear. Observations of the refracted waves are not nowadays used so much for local geological investigations because it is found that more detail can be obtained with greater economy by using reflected waves. One of the main problems with observing refracted waves is that the distance between the source and detector must be large compared with the depth of the layer in which the refracted wave travels and the expense and complexity of the observations increases with that distance. The way in which the time of arrival of a reflected ray depends upon the distance between the source and the seismometer can be obtained from the geometry shown in figure 3.15. If d is the distance between source and seismometer, and h the depth to the reflecting surface, the distance travelled by the reflected ray is

$$2(h^2 + \tfrac{1}{4}d^2)^{1/2}$$

and the time is therefore

$$t = \frac{2}{v} (h^2 + \tfrac{1}{4}d^2)^{1/2}$$

Thus

$$v^2 t^2 - d^2 = 4h^2$$

showing that the travel–time curve of a reflected ray is an hyperbola; the intercept upon the time axis is $2h/v$ and the asymptotes are the lines $t = \pm d/v$. Reflected waves returned almost normally from the interface can be observed close to the source and then any slope in the interface may be looked for by laying out

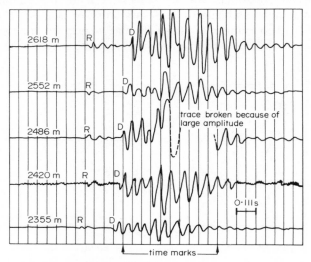

R: refracted wave from Palaeozoic floor
D: direct wave through surface rocks

Figure 3.13 Example of record of pulses from explosion in seismic survey in East Anglia (Bullard *et al.*, 1940). The distances in metres are those of the shot point from the seismometer

detectors in all directions around the source. The disadvantage of using reflected waves is that reflections often occur from boundaries within rocks that have very little significance. This is an example of a problem that always arises in investigating geological structures by geophysical methods—the physical properties of rocks, such as seismic velocity or density, do not necessarily correlate well with geological distinctions—two clays with quite distinct fauna lying unconformably one upon the other and readily distinguished by a geologist in the field may have almost identical densities or velocities, while within a geologically homogeneous bed there may be small variations of velocity giving rise to many reflections of seismic waves. Fortunately, these are problems that mostly affect small-scale and rather local studies and then, if the work is undertaken commercially, it will be checked from place to place by drilling boreholes.

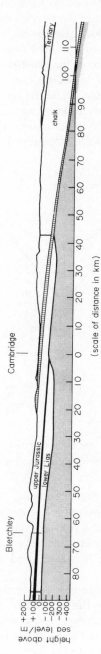

Figure 3.14 Structural section in East Anglia (Bullard *et al.*, 1940). Palaeozoic basement is shaded

Because of the difficulties reflected waves have until recently not been used to study the deep structure of the crust of the Earth, although now, at great expense, they are so used, especially in Russia; nonetheless most of our knowledge of the deep structure of the crust has been obtained from studies of refracted waves. The first indication of the structure of the crust came from the realization that surface irregularities were in isostatic equilibrium with compensating masses at depths less than 100 km (see chapter 2) and the idea of a relatively thin (50 km) less dense crust lying upon a denser material was developed. At much the same time, A. Mohorovičić interpreted his observations of the times of travel of pulses from some small earthquakes in Europe in terms of an upper layer with low seismic

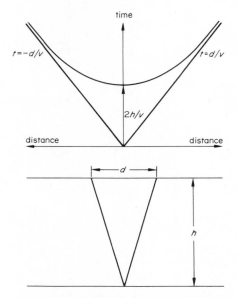

Figure 3.15 Travel–time curve for reflection

velocities resting upon one with a higher velocity. Details were filled in so far as possible from further observations upon pulses from small earthquakes, but since 1950 it has become possible as a regular matter to study the structure of the crust through observations of signals from large explosions, mainly on account of the development of improved seismometers and of methods for making observations at sea.

The methods used on land may be adapted to shallow water by placing geophones in watertight containers on the bed of the sea and connecting them to a recording ship through cables, and explosive charges may be placed upon the sea bed, the pressure of water above them ensuring that a substantial part of the energy of the explosion is coupled into the rocks of the sea floor. There remain problems of knowing where the source and the detectors are. In deep water it is not practical to put equipment on the bottom and present methods depend to a large extent on the work of M. N. Hill and his collaborators (Hill, 1966). They found that if a large explosion occurred in the sea even quite far above the floor adequate energy was transmitted from the wave in the water into the underlying rocks. The main reason for this is that the material lying on the sea bed is sand or mud with a very high proportion of water in the pores and with elastic properties not so very different from those of water. Similarly, a wave arriving at the sea floor from below generates an adequate wave in the water above, and that wave can be detected by an instrument adjusted to be sensitive to waves in water, that is, a hydrophone. Thus it is not necessary to place explosives or instruments on the sea bed, but they may be suspended at a convenient depth in the water. Hill also solved the problem of location by using the time of travel of the wave that goes direct from explosion to detector through the water as a measure of the distance between source and detector. The velocity of sound waves in

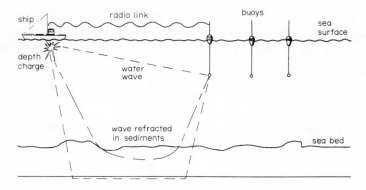

Figure 3.16 Layout of sono-radio buoys for marine seismic investigations

water is of course well known for a wide range of conditions. Hill suspended the hydrophones from free drifting buoys equipped with radio transmitters through which the signals from the hydrophones were transmitted back to the ship on which they were recorded and from which the explosives were set off. A recent improvement has been to provide the buoys with radar reflectors. A diagram showing the principle of the method is given in figure 3.16, while figure 3.17 shows a travel–time curve and corresponding structure.

The structure of the ocean floor is now well understood and, being simpler than that of the continents, is described first. Immediately below the water there is a layer of sediment of very high water content near the top and about 1 km thick in the Atlantic and 0·5 km in the Pacific. The water content decreases and the seismic velocity increases appreciably with depth as the sediment becomes more compact. P-wave velocities range from 1·5 to 4 km s^{-1}. The next layer down has a P-wave velocity of about 5 km s^{-1} and is some 1·7 km thick, and it is underlain by the main oceanic crust in which the P-wave velocity is 6·7 km s^{-1}. This in

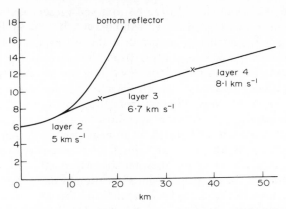

Figure 3.17 Typical oceanic travel–time curve. Ordinate, time in seconds

turn lies upon a material with a P-wave velocity of about $8 \cdot 1$ km s^{-1} known as the
upper mantle. By combining the thicknesses found from seismic data with gravity
observations made at sea, it is found that the densities of the layers under the
abyssal plains are typically

$$\text{layer 2: } 2700 \text{ kg m}^{-3}$$
$$\text{layer 3: } 3000 \text{ kg m}^{-3}$$
$$\text{layer 4: } 3400 \text{ kg m}^{-3}$$

Taking a representative depth of the oceans to be 5 km, the corresponding
thicknesses of the crustal layers are as shown in figure 3.18.

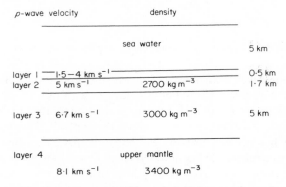

Figure 3.18 Representative section of the oceanic crust

The structure of the continental crust shows greater variations from place to
place and is not so clearly separated into distinct layers. In the simplest case, a
layer of uniform density overlies the upper mantle and the thickness of the crust
varies with the height of the land above the sea so as to maintain isostatic balance.
If the density of the upper part of the crust is 2700 kg m^{-3} and if that of the lower
part of the crust exceeds that of the upper mantle by 600 kg m^{-3}, then the load of

a mountain 2 km high will be compensated by a protrusion of the crust into the mantle equal to

$$\frac{2700 \times 2}{600} \text{ km}$$

that is, by 9 km. Thus, if the crust under land at sea level were 30 km thick, that under mountains with an average height of 2 km would be 39 km (below sea level). In some places the crust may be divided into two layers, the upper, less dense, of variable thicknesses, and the intermediate layer at the base of the crust that is denser but of more or less constant thickness. The Alps show a structure of this form, as indicated in figure 3.19. The boundary between the upper and inter-mediate layers is known as the Conrad discontinuity by analogy with the Mohoro-vičić discontinuity at the base of the crust, but it has nothing like the same status,

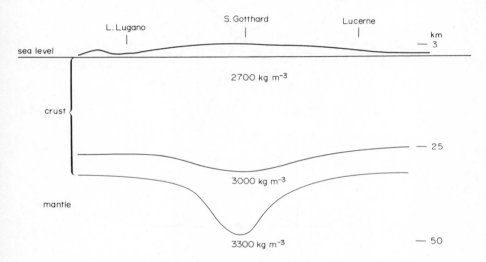

Figure 3.19 Structure of the Alps

for wherever observations have been made the Mohorovičić discontinuity defines the base of the oceanic as well as of the continental crust and separates layers of very distinct density and seismic velocity. The Conrad discontinuity, by contrast, is sometimes present and sometimes absent, it is a continental and not an oceanic feature and it separates materials which are often not sharply dis-tinguished.

Typical crustal material has a density of about 2700 kg m^{-3} compared with about 3000 kg m^{-3} for the main part of the oceanic crust, while the P-wave velocity is 6 km s^{-1} compared with 6·7 km s^{-1}. The surface layers comprise a complex set of rocks of more or less well-defined types made up of a selection of crystalline minerals, or in some cases, glasses, which are predominantly silicates.

The structure of the silicate minerals is based on the tetrahedral unit of one silicon ion surrounded by four oxygen ions (figure 3.20). Silicon has four valencies and oxygen has two so that, of the eight oxygen valencies, four are satisfied by the silicon valencies and there remain four to be satisfied in other ways. The ways in which the tetrahedral units are joined by other atoms to satisfy the remaining oxygen valencies determine the structures of the main types of silicate minerals.

Silica, SiO_2, consists of silica tetrahedra linked together by the oxygen atoms, the formula corresponding to $Si + 4(\frac{1}{2}O)$. In the olivines, on the other hand, the oxygen atoms of adjacent tetrahedra are linked by metal ions. Suppose that

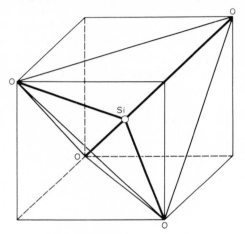

Figure 3.20 The silicate tetrahedron

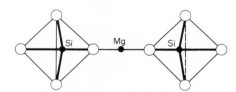

Figure 3.21 The structure of olivine

the metal M has two valencies one of which is linked to an oxygen ion in one tetrahedron and the other to an oxygen ion in an adjacent tetrahedron (figure 3.21). Four metal ions will be required to join the four silicon ions of one tetrahedron to adjacent tetrahedra and, since each metal ion is shared between two tetrahedra, the formula for olivines is $SiO_4 + 4(\frac{1}{2}M)$ or M_2SiO_4. The metal ions are either iron or magnesium and in general olivines consist of solid solutions having the formula $[xMg(2 - x)Fe]SiO_4$, where x may have any value between 0 and 2.

The iron olivine is called *fayalite* and the magnesian olivine, *forsterite*.

Between the extremes of silica with no metal ions joining the tetrahedra, and the olivines with the greatest possible number, there fall other families of minerals with the metal content increasing from silica to olivine. As with olivine, the sites for the metals in the crystal lattice may be occupied by various different metal ions, and so each main group of minerals comprises a range of solid solutions in which the relative proportions of the metals vary continuously. Counting silicon as a metal, the ratio of metal to oxygen is $1:2$ in silica and is $3:4$ in olivine. The minerals with the ratio $5:8$ are called *feldspars* and have the general formula $M_2AlSi_2O_8$, where M stands for two metal ions of which one may be potassium,

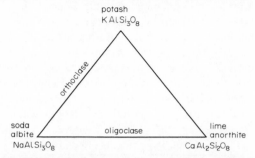

Figure 3.22 Triangle diagram of the composition of feldspars

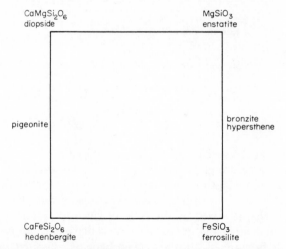

Figure 3.23 Rectangle diagram—the composition of pyroxenes

sodium or calcium, while the other will be aluminium or silicon. The way in which the metals can replace each other is shown in the triangle diagrams of figure 3.22. The minerals with the metal–oxygen ratio of $2:3$ are known as *pyroxenes* and have the formula $M_2Si_2O_6$, where the metals may be chosen from iron, calcium or magnesium or, more rarely, sodium, potassium, aluminium or titanium. The rectangle diagram of figure 3.23 shows the main elements of the pyroxene series. The densities of some representative minerals are given in table 3.1.

At the surface of the Earth we see rocks that have been formed at high temperature or pressure and others derived from them by the action of wind and rain, sunshine and frost; the first are known as igneous and metamorphic rocks and the second as sedimentary rocks, because they are found to have been deposited in more or less horizontal layers, or beds, by wind or water. Sedimentary rocks are now being formed, for example in deep trenches off the continental shelves or in deserts or in lakes and they comprise mostly silica derived from the mechanical breakdown of other rocks and clay minerals formed from feldspars through chemical action. The red colour of many sandstones comes from iron oxides derived from minerals containing iron. Another group of sedimentary rocks are those composed of calcium carbonate that have been formed from the shells of marine animals such as corals or oysters; they are the limestones.

Igneous rocks were at one time molten or effectively molten and the minerals found in them depend on the overall chemical composition of the rock and on the rate at which the molten material crystallized. Rocks which were molten when

Table 3.1   Densities of minerals

		Density $(kg\ m^{-3})$
Silica	quartz, SiO_2	2654
Feldspar	albite $NaAlSi_3O_8$	2620
	anorthite $CaAl_2Si_2O_8$	2760
	potash $KAlSi_3O_8$	2500
Pyroxene	diopside $CaMgSi_2O_6$	3300
	enstatite $MgSiO_3$	3270
	hedenbergite $CaFeSi_2O_6$	3550
Olivine	fayalite Fe_2SiO_4	4100
	fosterite Mg_2SiO_4	3200
Mica		2700–3150

they were ejected at the surface of the Earth from volcanoes cooled quickly and have small crystals or are glasses; they are called volcanic rocks. Rocks which cooled slowly at some depth in the Earth have large crystals and are known as plutonic rocks. Other rocks of immediate types also occur. The composition of igneous rocks is characterized by the ratio of silica to metals in the rocks as a whole. Rocks with a large amount of silica are called *acidic*, or *unsaturated*, while those with small amounts of silica are called *basic* or *saturated*. Granite is a plutonic rock composed of large crystals of silica and feldspars with small amounts of the mineral mica and is an acidic rock. Gabbro, on the other hand, is also a plutonic rock, but contains little or no free silica and is composed mainly of feldspar and pyroxene. Rhyolite is an acidic volcanic rock and basalt is a basic volcanic rock. There are also rocks having compositions intermediate between the acidic and basic, in which feldspars are the dominant minerals; diorite is an intermediate plutonic rock, andesite an intermediate volcanic rock. Rocks with less silica than basalt and gabbro consist only of pyroxenes and olivines and are called ultrabasic; peridotite is a mixture of pyroxene and olivine and dunite is an olivine alone. The ultrabasic rocks are plutonic, and are found in unusual sites such as diamond pipes or are dredged up from the floor of the deep ocean.

Metamorphic rocks are somewhat similar to plutonic rocks but have been formed by the recrystallization of pre-existing igneous or sedimentary rocks at high pressure and temperature, though without melting taking place. Metamorphic rocks are often found on the sites of old eroded mountains where they were subject to high stress and temperature during the formation of the mountains. Gneisses are metamorphic rocks of much the same composition as granite, while schists have been formed from clay rocks such as shales and slates.

The continents are generally thought to consist of a crust of granite or intermediate composition upon which lies a rather thin deposit of sedimentary deposits. The oceanic crust, on the other hand, is generally basaltic. The mantle which underlies the crustal layers is more ultrabasic and is commonly thought to consist in the main of olivine.

Table 3.2 Comparison of oceanic and continental crust

Oceanic				Continental			
Layer	Thickness (km)	Density (kg m^{-3})	Compressional velocity (km s^{-1})	Layer	Thickness (km)	Density (kg m^{-3})	Compressional velocity (km s^{-1})
water	5	1030					
sediment	0·5–1	1030–2700	1·5–4	upper crust	12	2700	5·8–6·4
consolidated sediment	1·5	2700	5				
basalt	5	2900	6·7	lower crust	18	2950	6·6–7·0
upper mantle		3310	8·1	upper mantle		3310	8

Note. The boundary between the upper and lower crust is very variable and may often not occur.

In certain areas the crust is found to show significant lateral variations. Extensive measurements in the U.S.A. are summarized in figure 3.24, from which it will be seen that the velocity and density of the crust (the latter found from the seismically determined thickness combined with gravity measurements) vary from place to place (figure 3.24(a)) as does the seismic velocity in the uppermost part of the mantle (figure 3.24(b)). Isostatic balance now implies differences of density within the crust as well as differences of the thickness of the crust.

Representative sections of oceanic and continental crust are summarized in table 3.2 which shows how isostatic balance occurs (see example 3.4).

While the oceans as a whole seem to be underlain by very uniform structures, there are some local structures of great interest and importance where the layers of the crust differ considerably from those normal over most of the oceans. The transition between continental and oceanic crust does not commonly take place smoothly but there is often a substantial thickness of relatively light mud or sand lying over the edge of the continent because it is there that sediment washed

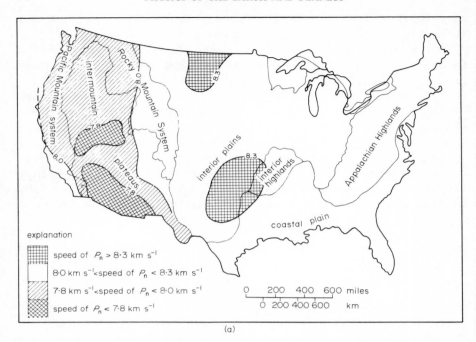

(a)

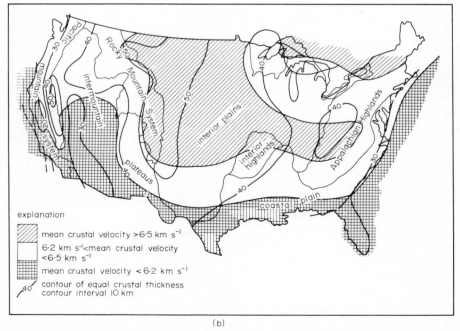

(b)

Figure 3.24 Crustal structure below the U.S.A.: (a) *P*-wave velocity at the top of the mantle; (b) *P*-wave velocity and thickness of the crust

down from the land accumulates. Two typical margin zones are indicated in the section in figure 3.25. As in other investigations of the crust of the Earth, the thicknesses of layers have been found from seismic observations and the densities of the layers are then calculated so as to reproduce the observed attraction of gravity.

Another important group of structures is the island arcs, of which well-known examples are the West Indies and the East Indies; they consist of chains of islands on a ridge in the sea bed in the form of an arc with a deep trough in the

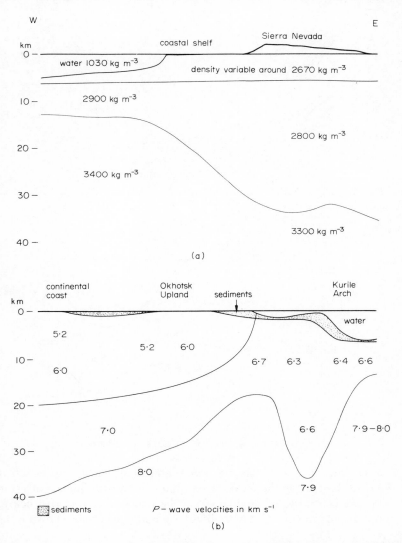

Figure 3.25 Structures of continental margins: (a) off California; (b) region of Okhotsk

sea bed of similar form lying just outside the chain of islands There is often
intense volcanic activity and a great thickness of sediments may accumulate.
It is usually thought that an island arc is in process of developing into a range of
mountains. The structure of the crust in an island arc is shown in figure 3.26.

The most imposing of all the superficial structures of the Earth is the system of

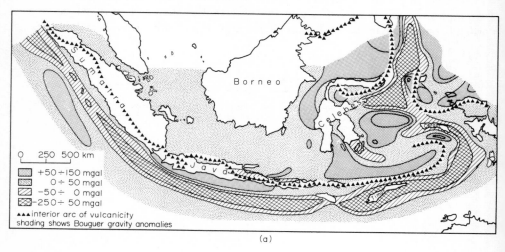

(a)

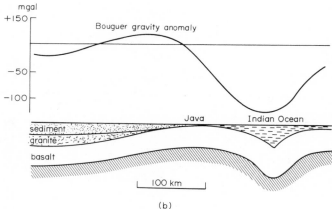

(b)

Figure 3.26 Map and structure of an island arc

mid-oceanic ridges that encircles the Earth and of which a map is shown in
figure 3.27. At these ridges the sea bed rises quite steeply to culminate in a summit
that is cut by a central valley (figure 3.28). The structure of the crust below the
ridge is not at present fully understood but a possible section is shown in figure
3.29.

The sections that illustrate different structures are all to a greater or lesser extent uncertain. Even when the depths to various layers are well defined from seismic data, there is often scope for a range of densities that will satisfy the observed attraction of gravity. In addition, interpretations of seismic data are not always clear cut, especially when slopes are appreciable or when, as may happen,

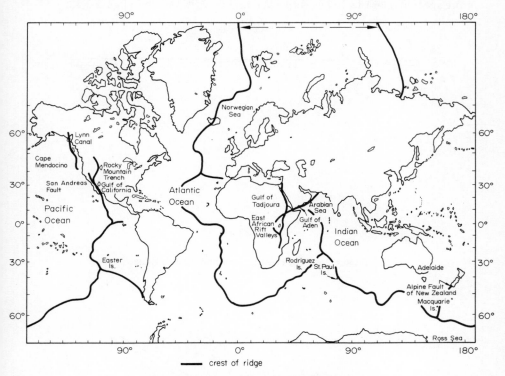

Figure 3.27 Map of the mid-oceanic ridges

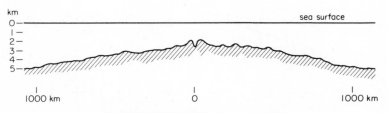

Figure 3.28 Profile of mid-oceanic ridge

it is not possible to get all the data needed to check on slopes. The same structural patterns are, however, often found to satisfy the observations in widely separated areas and so some confidence may be drawn from the fact that they apply on a world-wide basis.

The density of a rock and the velocity of seismic waves within it are both determined by the minerals of which it is composed and by the degree to which they are compacted, so that it is reasonable to look for some correlation between the two properties. In a general way, the harder a rock, the denser it is and the faster seismic waves travel through it. Many data are now available to enable more specific relations to be established and two are shown in figure 3.30.

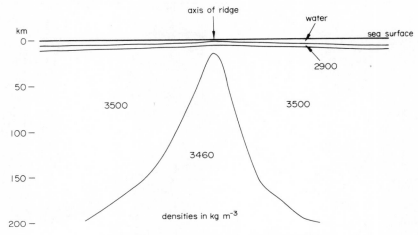

Figure 3.29 The crust below a mid-oceanic ridge

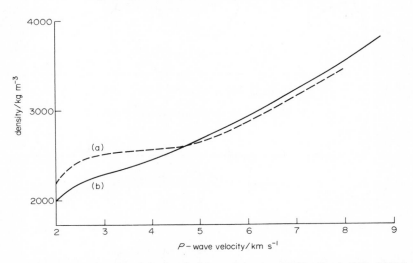

Figure 3.30 Relations between seismic velocities and density: (a) Woolland, 1959; (b) Talwani *et al.*, 1959

3.5 The deep interior of the Earth

The study of the deep interior of the Earth is based on a model in which the density and elastic properties depend on the radius but do not vary laterally.

If then an elastic disturbance travels between a source and a detector, symmetry shows that the ray path that it follows must lie in the plane containing the source, the detector and the centre of the Earth.

Surfaces on which the seismic velocity is constant are spheres concentric with the Earth and they therefore intersect any ray plane in circles concentric with the Earth. Consider, then, the path of a ray lying in a plane with the seismic velocity dependent only on the distance from the centre (figure 3.31).

Let O be the centre of the Earth and let the two shells S_1 and S_2 separate layers with the respective velocities v_1, v_2 and v_3. Let a ray making an angle of incidence i_1 with the radius vector OP_1 be refracted at P_1 and make an angle j_1 with OP_1

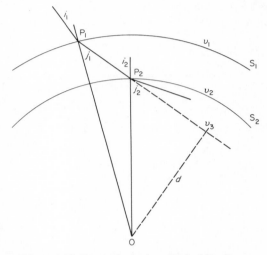

Figure 3.31 Ray paths in the spherical Earth

in medium 2. Let the corresponding angles for refraction at the point P_2 of S be i_2 and j_2. Then

$$\frac{\sin i_1}{\sin j_1} = \frac{v_1}{v_2}$$

so that

$$\frac{OP_1 \sin i_1}{v_1} = \frac{OP_1 \sin j_1}{v_2} \tag{3.2}$$

Now, if d is the perpendicular distance from O to P_1P_2,

$$d = OP_1 \sin j_1 = OP_2 \sin i_2$$

and therefore

$$\frac{OP_1 \sin i_1}{v_1} = \frac{OP_2 \sin i_2}{v_2} \tag{3.3}$$

It follows that $p = (r \sin i)/v$ is a constant along the ray (r is the radial distance from the centre).

p is called the *parameter* of the ray. At the deepest point reached by the ray, the angle of incidence i is $\frac{1}{2}\pi$ and so

$$p = r_{\min}/v$$

where the value of $r_{\min}$ is the radius of closest approach to the centre and v is the velocity at that point.

Now consider two adjacent rays symmetrical about the same radius vector, as shown in figure 3.32. Let the termini of the shorter ray subtend an angle Δ at the centre and let the time taken to travel along the ray be T. Let the angle and time for the longer ray be $\Delta + \delta\Delta$ and $T + \delta T$ respectively. Let S_1 and S_2 be

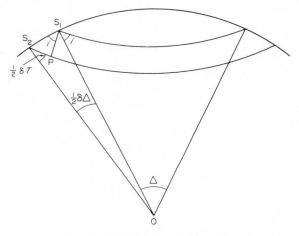

Figure 3.32 Geometry of adjacent rays

the origins of the shorter and longer rays respectively, and let P be the foot of the normal dropped from S_1 on to the longer ray. If i is the angle between the ray through S_1 and the radius OS_1,

$$\sin i = \frac{S_2P}{S_1S_2}$$

But $S_2P = v_0\frac{1}{2}\delta T$ and $S_1S_2 = r_0\frac{1}{2}\delta\Delta$, where v_0 and r_0 are the surface velocity and radius. Hence

$$p = \frac{r_0 \sin i}{v_0} = \frac{\partial T}{\partial \Delta} \tag{3.4}$$

The fact that p is equal to $r_{\min}/v$ enables v to be found as a function of r if p is derived by this formula from T and Δ. The theory is given in appendix 9.

Tables giving the times of travel of P and S pulses from earthquakes were first compiled in 1900 by Oldham and subsequently by Zöppritz. A very thorough analysis of the times of earthquake pulses was carried out by Sir Harold Jeffreys in collaboration with K. E. Bullen, and resulted in the Jeffreys–Bullen tables of 1940 which were the standard to which all studies of earthquakes were referred. The travel times of pulses from earthquakes are subject to a number of uncertainties of which the most important are the real variations of velocity from place to place and the errors that arise because the position and time of an earthquake are generally unknown and so have to be found from the observed times. Nothing

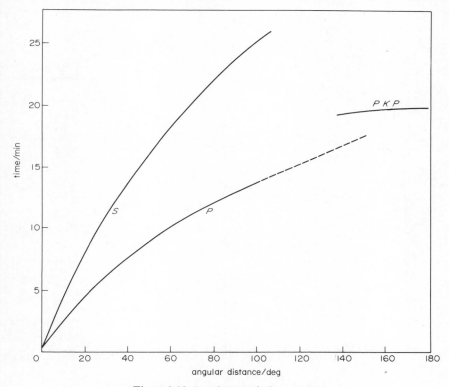

Figure 3.33 *P* and *S* travel–time curves

can be done about the real variations of properties but the scatter due to ignorance of the source has been very greatly reduced in observations of pulses from large nuclear explosions for which the times and places of the explosion are generally well determined. Not only are the geographical coordinates well known but it is certain that the sources are at the surface of the Earth whereas the fact that the depth of a natural earthquake focus is also initially unknown may give rise to further uncertainty. Not surprisingly, much better travel–time tables can now be produced using data from nuclear explosions. The present standard tables are those of Herrin and his collaborators (Herrin, 1968).

Given a set of tables for the times of the *P* and the *S* waves that travel directly between two points at the surface, the distribution of the corresponding velocities within the outer parts of the Earth may be derived. Figure 3.33 shows the travel-time curves for *P* and *S* waves, and figure 3.34 the corresponding velocity curves. If the velocities are known, then the times of travel for other possible ray paths may be calculated, for example, the time of a ray that, as shown in figure 3.35 is reflected one or more times at the outer surface. Reflected rays may also suffer transformation from *P* to *S* waves or *vice versa*. Similarly the effect of the

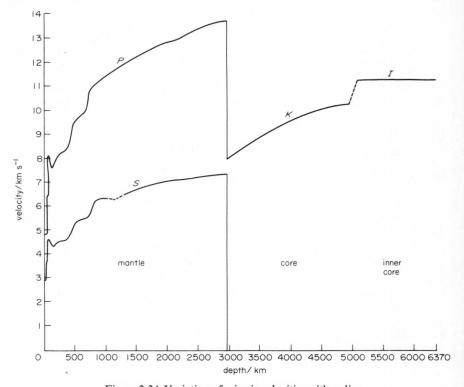

Figure 3.34 Variation of seismic velocities with radius

depth of the earthquake upon the time may be calculated, both for rays going directly to the surface as well as for those that are first reflected at the surface. When predicted times for the various rays are available, the rays can be identified in earthquake records and then themselves used to improve the values of the velocities. Figure 3.35 shows some of the ray paths and figure 3.36 the corresponding travel–time curves.

The foregoing methods are valid provided the velocities vary continuously with radius, for then the times and amplitudes of the seismic pulses would change steadily as the distance from the source increases; sometimes, however, they are found not to do so. At angular distances from about 105° to 140° the amplitudes of P waves are very greatly reduced, but at 143° the amplitude is large, as shown in figure 3.37, and at greater distances two P pulses occur. These facts can be explained by supposing that at a radius of about 3470 km the P velocity decreases sharply. The value of Δ for a ray tangential to the sphere of that radius is about 105° and no ray with a larger value of Δ can reach the surface directly. The region within 3470 km is known as the *core* of the Earth and, because rays that reach it are refracted in towards the radius vector, it causes a shadow that extends from 105° to 140°. Just beyond 140°, however, the core focuses waves into a caustic,

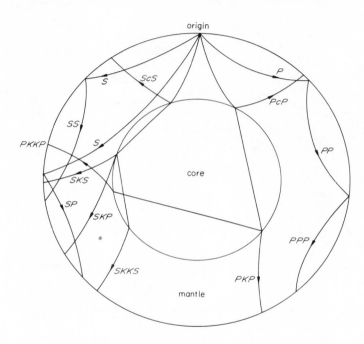

Figure 3.35 Composite ray paths in the Earth

so producing the large amplitudes. The waves that pass through the core are denoted by K and, if they start off and finish up as P waves in the outer region, they are labelled *PKP*. Figure 3.38 is a diagram illustrating the formation of the shadow zone and the caustic. The wave K is a dilatational wave and no signals have been observed that would correspond to shear waves through the core; it is therefore concluded that the core cannot transmit shear waves and is liquid.

Since the velocities of waves outside the core are known, travel–time curves for the ray K starting and ending on the surface of the core may be derived from the surface travel–time curves for the *PKP* rays. The distribution of velocity within the core can then be found from such a derived travel–time curve just as

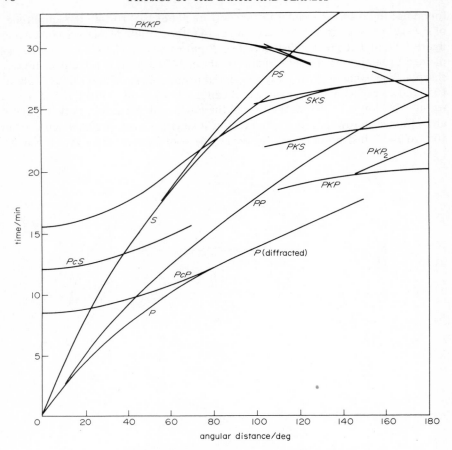

Figure 3.36 Travel–time curves for composite paths

Figure 3.37 A seismogram at a distance greater than 140°

the velocity within the upper parts of the Earth is found from the actual P travel–time curve. Figure 3.35 shows PKP and other rays associated with the core and figure 3.34 shows the variation of velocity within the core. The innermost part of the core is considered to be solid (see chapter 5).

The travel–time curve for P waves shows a discontinuity in slope at values

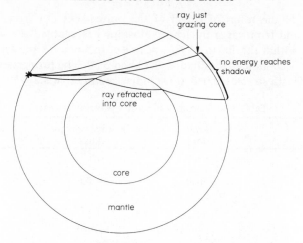

Figure 3.38 Formation of the shadow zone

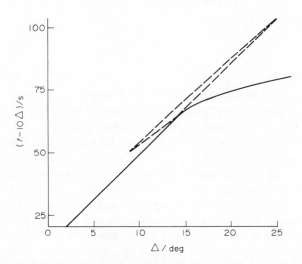

Figure 3.39 Cusps in travel–time curves corresponding to discontinuous increase of velocity at 200 km

of Δ near 20°, as does that for S waves. Calculations show that, if the velocity increases more rapidly between the depths of about 600 and 900 km than elsewhere, then the travel–time curve should show cusps as indicated in figure 3.39. The observed behaviour corresponds well to such a model and the velocities of the P and S waves are found to vary as shown in figure 3.34. The discontinuity is known as the 20° discontinuity and the zone of more rapid increase of velocities separates the *upper mantle*, which extends to the crust, from the *lower mantle*, which stretches down to the core.

The study of the times of travel of the pulses sent out from large natural earthquakes and from large nuclear explosions has enabled the velocities of P and S waves within the Earth to be traced out in some detail, with results that have already been indicated in figures 3.33 and 3.34. The properties of the major zones of the Earth so established are summarized in table 3.3.

Table 3.3   Velocities of seismic waves in the earth

Zone	Depth (km)	P-wave velocity (km s^{-1})	S-wave velocity (km s^{-1})
upper mantle	6300	7·8	4·4
	5600	10·6	5·8
lower mantle		13·6	7·3
	3500		
		8·10	0
core		9·40	0
	1250		
		11·2	2
inner core			
	0	11·3	2

3.6 Surface waves

Waves which travel by the various paths through the Earth that have been described in the previous section give rise to quite sharply defined movements of the ground. The seismometer response usually appears as a few cycles of a periodic signal, starting abruptly within a fraction of a cycle. There is usually a small background of more or less periodic motion generated by waves which have been scattered at other discontinuities than the major ones, but the principal signals are generally clear, especially at the greater distances from the source. The impulsive signals are followed by a long train of periodic disturbances of greater amplitude than the impulsive signals and continuing often for hours; it is known as the L phase and is caused by waves that have travelled much more slowly than the waves through the body of the Earth and contain much more energy, both because they have larger amplitudes and because of the much greater time for which they continue. These are the characteristics of waves that have been guided along a surface.

Waves guided along surfaces are familiar in other fields of physics. Sound waves in the sea can be guided over very great distances by the surface of discontinuity between two layers of water of different density, while electromagnetic waves can be guided by pairs of parallel conductors, by hollow tubes (waveguides), by dielectric rods and by the conducting surface of the Earth. Guided waves are characterized by boundary conditions that must be satisfied at the guiding surface and by the disturbance dying away more or less exponentially from the surface. Because the wave motion is confined to the neighbourhood of the surface, the area of the front of a wave spreading out from a point source is proportional to the distance from the source, so that the intensity of the wave motion is inversely proportional to the distance and the amplitude inversely proportional to the square root of the distance. On the other hand, when body waves spread out from a point source throughout the entire volume of a solid, the area of the wave front is proportional to the square of the distance from the source, the intensity is inversely proportional to the square of the distance and the amplitude inversely proportional to the distance. It is this geometrical fact that explains the much greater amplitude of the L phase of a seismometer record at great distances from a source.

There are essentially two types of surface wave that can be propagated over the surface of the Earth: the one, first studied by Lord Rayleigh (1885) and known as *Rayleigh waves* can be supported at the surface of a uniform semi-infinite medium, whilst the other, known as *Love waves*, after A. E. H. Love (1911), can only be supported by a relatively thin layer on top of a semi-infinite solid, a situation corresponding to the crust of the Earth lying on top of the mantle. The theory is given in appendix 10.

Because the speed of Love waves depends on the period, a signal that starts as a sharp pulse containing a wide range of periods will be spread out in time when it arrives at a distant observer, energy at different periods having taken different times to make the passage. The original pulse is therefore dispersed into a train of periodic waves, the dominant period changing with time of arrival, and, for this reason, waves of which the speed depends on period are called *dispersive*. Gravity waves and surface tension waves on water are other examples of dispersive waves.

Since the expression for the displacement in Love waves involves the phase factor $i(\kappa x - vt)$, it can be seen that the velocity C of a group of waves with speed v is

$$\partial v / \partial \kappa$$

Because v/κ is equal to the phase velocity, c,

$$C = \frac{\partial(\kappa c)}{\partial \kappa} = c + \kappa \frac{\partial c}{\partial \kappa}$$

The behaviour of c and C for Love waves is indicated in figure 3.40.

Rayleigh waves on a semi-infinite medium are not dispersive but, because the Earth is layered, their behaviour on a layered system must be considered. The theory is much more complicated than that of Love waves on two layers or Rayleigh waves on a semi-infinite medium and involves the solution of a determinant of sixth order to find the speed. The waves are dispersive and the behaviour of the phase and group velocity is indicated in figure 3.41.

The general features of Love waves and Rayleigh waves detected on the Earth can now be summarized. Such waves will travel to the detector along a great circle path from the source, and will be most strongly excited by a source at the surface where the amplitude of the wave motion is greatest. The first waves will

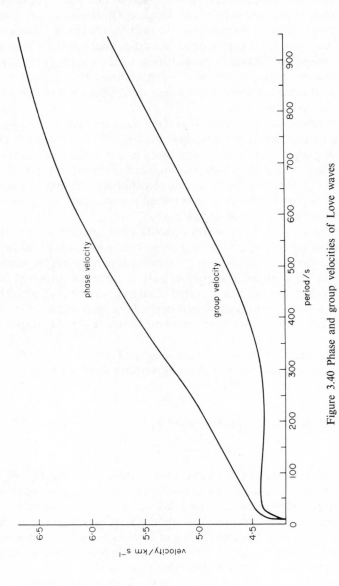

Figure 3.40 Phase and group velocities of Love waves

have a very well-defined period, that for the greatest group velocity, and as time goes on will be succeeded by those of smaller velocity. In favourable circumstances it is possible to detect waves that have passed more than once round the Earth. Love waves may be distinguished from Rayleigh waves by the fact that they have no vertical component.

Given the time of arrival of waves of a given period, a curve of group velocity against period can be constructed and used to examine the structure of the outer parts of the Earth. The waves of short wavelength, say 10 km, are most sensitive to variations within the crust of the Earth and for them it is possible to ignore

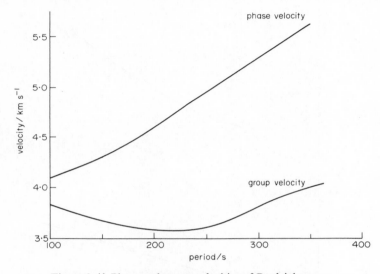

Figure 3.41 Phase and group velocities of Rayleigh waves

the curvature of the Earth. The main factor that affects the transmission of waves of short wavelength is the gross difference between the crust of the continents and that of the oceans. If the wavelength is long compared with the thickness of the crust, the waves are not much affected by the properties of the crust but the transmission is now controlled by the properties of the upper mantle, while the effect of curvature can no longer be ignored.

The equation for the speed of Love waves has many solutions, not just a single one, the different solutions corresponding to the number of times that the amplitude of the wave vanishes on nodal planes within the upper medium. Such higher mode solutions are not of much importance within the crust but do arise in considering waves transmitted in the mantle.

The equations of motion for Rayleigh and Love waves were set up and solved with the implicit assumption that the density and elastic moduli within a layer were constants. In practice they are not and the equations must be formulated to take the variations of those quantities explicitly taken into account. The equations of motion still separate into equations that are satisfied by waves of Rayleigh and Love type, but they cannot be solved analytically. The general procedure therefore is to calculate the dispersion curves for the two types of wave for Earth models with various assumed variations of density and elastic moduli and to compare the results with the observed dispersion curve, adjusting the parameters of the model until reasonable agreement is attained. In this way it has been found that the S, and possibly the P, wave velocity has a minimum around 200 km (figure 3.42).

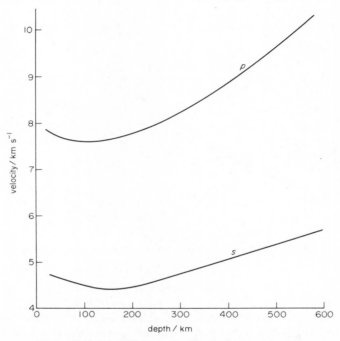

Figure 3.42 P and S wave velocities in the upper mantle

Examples for chapter 3

3.1 Calculate the bulk modulus and shear modulus of a material having the following properties:

density	4000 kg m^{-3}
dilatational velocity	10 km s^{-1}
shear wave velocity	6 km s^{-1}

3.2 The elastic wave velocities in two media are respectively:

medium 1: $\alpha = 6$ km s^{-1} medium 2: $\alpha = 8$ km s^{-1}
$\quad\quad\quad\quad \beta = 4$ km s^{-1} $\quad\quad\quad\quad\quad\quad \beta = 5$ km s^{-1}

Calculate the angles of reflection and refraction for both types of wave when the incident wave is dilatational and is incident at 48°36′.

3.3 Calculate the travel–time curve for arrivals up to 10 km of the first pulse from an explosion on the surface when the underlying structure consists of a layer of thickness 200 m in which the P-wave velocity is 5 km s^{-1} overlying material in which the velocity is 6 km s^{-1}.

3.4 In oceanic areas, the upper 50 km of the Earth consists of
    sea water, density 1030 kg m^{-3}, depth 5 km
    sediments, density 2500 kg m^{-3}, thickness 1 km
    crust, density 3000 kg m^{-3}, thickness 5 km
    upper mantle, density 3300 kg m^{-3}
In a continental area the same thickness is formed of
    upper crust, density 2700 kg m^{-3}, thickness T km
    lower crust, density 3000 kg m^{-3}, thickness $30 - T$ km
    upper mantle, density 3300 kg m^{-3}
Calculate T for isostatic equilibrium.

3.5 Calculate the parameter, p, for the following.
    (a) A ray that reaches a depth of $0.90R$, where R is the radius of the surface of the Earth, 6400 km.
    (b) A ray that reaches a depth of $0.90R_1$, where R_1 is the radius of the core, 3500 km.
Calculate $d\theta/ds$ and the angles of incidence for (a) at the surface of the Earth, and for (b) at the surface of the core.
Use the following velocities:
    at R: 7·75 km s^{-1} at $0.90R$: 10·55 km s^{-1}
    at R_1: 8·10 km s^{-1} at $0.90R_1$: 8·53 km s^{-1}.

3.6 Suppose that the upper parts of the Earth are composed of a series of layers of different seismic velocity, the thickness of the ith layer being h_i and the velocity in it being v_i, and with $v_{i+1} > v_i$. If the nth layer extends downwards indefinitely, show that the time taken for a ray to travel from a source S down to the nth layer, along the nth layer and back up to a detector D at a distance L is

$$\frac{L}{v_n} + 2 \sum_{i=1}^{n-1} h_i \left(\frac{1}{v_i^2} - \frac{1}{v_n^2} \right)^{1/2}$$

Further reading for chapter 3

Bullard, E. C., Gaskell, T. F., Harland, W. B., and Kerr-Grant, C., 1940, 'Seismic investigations on the Palaeozoic floor of East England', *Phil. Trans. Roy. Soc.*, *Ser. A*, **239**, 29–94.

Bullen, K. E., 1963, *Seismology* (Cambridge: Cambridge University Press).

Herrin, E., 1968, '1968 seismological tables for P phases', *Bull. Seismol. Soc. Am.*, **58**, 1193–1241.

Hill, M. N. (Ed.), 1966, *The Seas*, Vol. III (London: Interscience), xvi + 963 pp.

Jeffreys, H., 1970, *The Earth*, 5th edn (Cambridge: Cambridge University Press), xii + 525 pp.

Talwani, M., Sutton, H., and Worzel, J. L., 1959, *J. Geophys. Res.*, **64**, 1545.

Tucker, R. H., Cook, A. H., Iyer, H. M., and Stacey, F. D., 1970, *Global Geophysics* (London: English Universities Press), 199 pp.

Woollard, G. P., 1959, *J. Geophys. Res.*, **64**, 1521.

CHAPTER FOUR

Oscillations and vibrations

'The aged Earth aghast . . .
Shall from the surface to the centre shake.'

J. Milton, *Hymn on the morning of Christ's nativity*

4.1 Introduction

There are forces acting on the Earth that cause it to oscillate as a whole about its centre of mass and to vibrate internally. The attractions of the Sun and the Moon upon the equatorial bulge of the Earth vary with the positions of the Sun and the Moon and drive both oscillations of the Earth about its centre of mass (*precession* and *nutation*) and internal oscillations (*Earth tides*), while the shocks of the very largest earthquakes set the Earth into perceptible free vibrations. Each of these phenomena is dependent on the internal structure of the Earth and provides significant information about the distributions of density and elastic moduli within the Earth and also about the departures of the material of the Earth from purely elastic behaviour.

These oscillations and vibrations are well understood, either because they are free vibrations or because the driving force is well defined. There are other small motions of the Earth which are less well understood. Variations of the length of the day, that is, of the speed of rotation of the Earth, which may be regarded as small oscillations about the axis of rotation, are in part due to the attraction of the Sun and the Moon and in part due to seasonal variations in the moment of inertia of the Earth, but there are additional variations which cannot be ascribed to any well-defined cause. There are also small changes in the direction of the Earth's axis of rotation that are still quite obscure.

It has not so far been possible to look for tidal oscillations or free vibrations of the Moon but the varying attraction of the Earth and the Sun upon the Moon result in oscillations about its centre of mass known as *physical librations*; they provide information about the moments of inertia of the Moon.

4.2 Precession and nutation

To a close approximation, the Sun and the Moon may be considered to move around the Earth in the same plane, the *ecliptic*, that is the plane of the Earth's orbit about the Sun (the inclination of the Moon's orbit to the ecliptic is 5·9° and it is because it is so small that eclipses occur quite frequently). The ecliptic is inclined at 23°27′ to the equator of the Earth and so, as the Sun and the Moon move in their respective orbits, they appear at times to the north, at times to the south of the equator and thus the torques they exert on the Earth, proportional to the attractions they exert on the equatorial bulge, vary periodically, in the one case with an annual period, in the other with a monthly period. On the average, the torque is exerted about the line of intersection of the equator and the ecliptic and, because the Earth is spinning about the axis perpendicular to the equator, gyroscopic torques cause the Earth to oscillate about an axis perpendicular to those two directions. The gyroscopic motion has a steady component, the *precession*, on which is superposed the periodic *nutation*. As shown in appendix 11, the intersection of the equator and the ecliptic moves backwards along the equator at the steady rate $-\frac{3}{2}k\cos\theta_0$, where θ_0 is the mean inclination of the equator to the ecliptic and k is $GM'(C-A)/C\tilde{\omega}r^3$. M' is the mass of the Sun or the Moon, $\tilde{\omega}$ is the spin angular velocity of the Earth and r is the distance of the Sun or the Moon from the Earth.

The Sun and the Moon both contribute to the precession, the total rate being $\frac{3}{2}(k_S + k_M)\cos\theta_0$, where k_S and k_M are the solar and lunar factors. The arguments of the solar and lunar nutations are different, for n is in the one case the orbital speed of the Sun (2π in 1 year) and in the other case it is that of the Moon (2π in 1 month). The nutation therefore shows oscillations with periods of half a year and a fortnight. k_S is $(GM_S/\tilde{\omega}r_S^2)\{(C-A)/C\}$, M_S being the mass of the Sun and r_S its distance.

By Kepler's law, if n_S is the orbital angular velocity of the Sun,

$$n_S^2 r_S^3 = GM_S$$

and so

$$k_S = \tilde{\omega}\left(\frac{n_S}{\tilde{\omega}}\right)^2 H \tag{4.1}$$

where H is the usual abbreviation for $(C-A)/C$. H is often called the *dynamical ellipticity* of the Earth.

Because the Moon is much less massive than the Earth, its orbital angular velocity satisfies

$$n_M^2 r_M^3 = GM_E$$

M_E being the mass of the Earth. Thus

$$k_M = \tilde{\omega}\left(\frac{n_M}{\tilde{\omega}}\right)^2 \frac{M_M}{M_E} H \tag{4.2}$$

Now $n_S/\tilde{\omega} = 1/365$ and $n_M/\tilde{\omega} = 1/28$, while H is about $1/300$ and M_M/M_E is $1/81$. Thus

$$k_S/\tilde{\omega} = 9 \times 10^{-6} \text{ rev y}^{-1}$$

and

$$k_M/\tilde{\omega} = 1.9 \times 10^{-5} \text{ rev y}^{-1}$$

giving for the net rate, $\frac{3}{2}(k_M + k_S)\cos\theta_0$, the value 3.86×10^{-5} rev y^{-1}.

The rate of precession is known very accurately from observations of the changes in the apparent positions of the stars but, to calculate H with high accuracy, attention has to be paid to the following points.

First, a number of simplifications have been made in the foregoing outline of the theory, especially by neglecting the eccentricities and inclinations of the various orbits.

Secondly, a very accurate value of M_M/M_E is needed since k_M is greater than k_S—in fact, the ratio is now known very accurately from a variety of space probe measurements.

Thirdly, the planets also contribute to the overall precession and there is a small effect of general relativity.

When these refinements are taken into account, it is found that

$$H = 3 \cdot 273 \times 10^{-3}$$

4.3 The librations of the Moon

The Moon, as is well known, spins upon her axis at the same rate as she goes round the Earth in her orbit and so presents, on the average, the same face to the Earth. Because the orbit is slightly eccentric and because it is inclined both to the ecliptic and to the equator of the Earth, the aspect of the Moon as seen from the Earth undergoes slight changes, the *optical* or *geometrical librations*, but in addition there are physical rotations of the Moon about its mean position which are called the *physical librations* and arise from the attraction of the Earth, and to a much less extent the Sun, upon the equatorial bulge and elliptical equator of the Moon.

The Moon, with an elliptical equator, has three distinct moments of inertia (A, B, C). She moves round the Earth in an orbit that is inclined at some 5° to the ecliptic and the principal axis about which the moment of inertia is greatest (C) is inclined at about $1 \cdot 5°$ to the pole of the orbit (figure 4.1). The axis about which the moment of inertia is least points in the mean direction of the Earth and the third axis is therefore tangential to the orbit of the Moon. This disposition of axes is the only stable one.

The direction of the Earth from the Moon changes periodically because the orbit of the Moon is elliptical and the Earth is sometimes above the lunar equator and sometimes below, so that couples are exerted about the axes in the plane of the equator. The theory of the librations is considerably more complex than that of the luni-solar precession of the Earth and will not be developed here (see, for example, Plummer, 1960). The main result is that the direction of the greatest principal axis of inertia of the Moon (C) moves around the pole of the orbit in an ellipse which is very nearly a circle with angular radius

$$\frac{3\beta \sin i}{(1 + \mu)(2p/n - 3\beta)}$$

and with angular velocity $n + p$.

Here $\beta = (C - A)/B$, i is the inclination of the Earth's orbit to the mean plane of the Moon's equator, μ is the ratio of the mass of the Moon to the mass of the Earth and n is the mean angular velocity of the Moon in her orbit.

p is the rate at which the node of the Moon's orbit regresses along the ecliptic (mainly on account of the attraction of the Sun).

$$p = 2\pi/18 \cdot 6 \, \text{y}^{-1}, \quad n = 2\pi/27 \cdot 3 \, \text{d}^{-1}, \quad p/n = 4 \times 10^{-3}$$

and $\beta = 6 \cdot 25 \times 10^{-4}$, so that $2p/n - 3\beta = 6 \times 10^{-3}$; thus there is an amplification of the motion of the pole of the Moon, the angular amplitude of which is about $1°32'$.

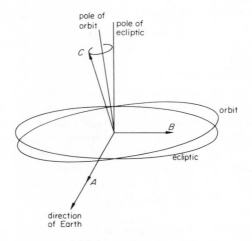

Figure 4.1 The Moon in relation to the Earth

The measurement of the librations has hitherto depended on the observation of the distance from the limb to a small crater, Mösting A, near the centre. The difficulties in the way of precise observations are very great and the data have been extensively discussed without attaining really reliable results. Recently, however, a new, more direct way of determining the librations has been made possible. The *Apollo 11* and later astronauts and the Lunokhod automatic exploration vehicle have placed reflectors on the Moon, with the aid of which it is proving possible to determine the time of flight of an optical laser beam from a point on the Earth to the lunar reflector and back.

Let R be the vector separation of the centres of mass of the Earth and the Moon, and let r_1 be the position vector of the terrestrial observatory relative to the centre of mass of the Earth and r_2 that of the lunar reflector relative to the centre of the mass of the Moon. Then (figure 4.2) the distance D between terrestrial observatory and lunar reflector is

$$D = |R - r_1 - r_2|$$

which, since r_1 and r_2 are very much less than R, is very nearly

$$R - p_1 - p_2$$

where p_1 and p_2 are the respectively projections of r_1 and r_2 upon R.

p_1, the terrestrial projection, depends on the terrestrial coordinates and varies with the speed of the axial spin of the Earth. p_2 depends on the angle between R and r_2, that is, upon the librations. It has a monthly period and so can be separated from other components of the variations of D. It should in future be possible to make a much better determination of the libration than is at present possible.

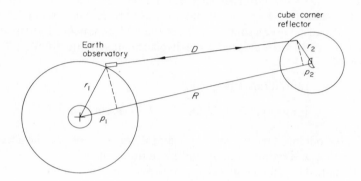

Figure 4.2 Laser ranging to the Moon

The motion depending on β is not the only libration; from a different motion it is possible to find γ, the ratio $(B - A)/C$. The present best values are

$$\beta : 6\cdot25 \times 10^{-4}$$
$$\gamma : 2\cdot32 \times 10^{-4}$$

4.4 The elastic yielding of the Earth as a whole

When the Earth is subject to a force throughout its body or to a load at the surface, it yields elastically. In general the forces to which the Earth is subject vary with time, as for example those due to the tidally varying attraction of the Sun and the Moon, but in some cases it is possible to consider the forces and strains to be static, whenever, in fact, the time taken for a disturbance to travel through the Earth is small compared with the period of the disturbing force. Taking that time to be of the order of 12 000 km divided by 15 km s^{-1}, or 800 s, it will be seen that tidal disturbances, with a period of half a day, can to a good approximation be regarded as static.

Ignoring then the terms $\rho(\ddot{u}, \ddot{v}, \ddot{w})$, in the equations of motion (see appendix 7), the equations for elastic equilibrium under body forces X, Y and Z per unit volume become

$$\left.\begin{array}{l} (\lambda + \mu)\dfrac{\partial \Delta}{\partial x} + \mu \nabla^2 u = X \\[2ex] (\lambda + \mu)\dfrac{\partial \Delta}{\partial y} + \mu \nabla^2 v = Y \\[2ex] (\lambda + \mu)\dfrac{\partial \Delta}{\partial z} + \mu \nabla^2 w = Z \end{array}\right\} \tag{4.3}$$

The only significant body forces are the tidal forces which, as will be seen, are proportional to second-order spherical harmonics, and so it is sufficient to consider the solution of the equations of equilibrium for a second harmonic body force. Because the equations are linear, they will be satisfied by displacements that are also second harmonics and the displacements at any point (r, θ, ϕ) may be written as

radially: $h_r V/g_r$

tangentially: $l_r \partial V/g_r \partial \theta$, $l_r \partial V/g_r \sin \theta \partial \phi$

where V is the potential from which X, Y and Z are derived, g_r is the acceleration due to gravity at the radius r and h_r and l_r are functions of r.

In particular, the radial displacement at the surface is written as

$$hV/g$$

and the tangential displacement as

$$l\partial V/g \, \partial \theta \quad \text{and} \quad l\partial V/g \sin \theta \, \partial \phi$$

The transfer of mass entailed by the displacements u, v and w gives rise to a change in the second harmonic part of the potential of the Earth which is written as

$$kV$$

h and k are called Love's numbers, having been introduced by A. E. H. Love. The use of l was suggested by W. Lambert.

The values of h, k and l, like the mass or moment of inertia, are integrals of the properties of the Earth throughout its volume and so provide a check upon distributions of properties rather than data from which the distributions can be calculated directly. Given a distribution of properties derived from seismic data, the Love numbers may be found by numerical integration of the equations of equilibrium.

The Earth may also be deformed by a surface load. A mass upon the surface of the Earth has two effects—the direct radial stress, $g\sigma$, where σ is the mass of the load per unit area of the surface, will depress the surface, while at the same time the additional potential of the load will produce displacements that will raise the surface. The equations of equilibrium can be solved in two steps: first the displacements due to the body forces corresponding to the potential of the load are found, and then to them are added the solutions without body forces that give the correct surface stresses. The results can be expressed in terms of Love numbers, but separate numbers are required for harmonics of different order.

Love numbers for harmonics beyond the second are only sketchily known and are not further considered. The remainder of this section is therefore concerned with the yielding of the Earth under the body force arising from the tidal potential of the Sun and the Moon.

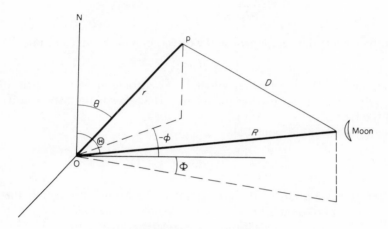

Figure 4.3 Geometry for tidal theory

Let the geocentric coordinates of the Moon be (R, Θ, Φ) (figure 4.3) and let those of a point P in the Earth be (r, θ, ϕ). The distance of the Moon from P is

$$D = (R^2 + r^2 - 2Rr \cos \chi)^{1/2}$$

where χ is the angle between the radii vectors R and r:

$$\cos \chi = \cos \Theta \cos \theta + \sin \Theta \sin \theta \cos (\Phi - \phi)$$

The potential of the Moon at P is then

$$V_{\mathrm{M}} = - \frac{GM_{\mathrm{M}}}{D}$$

where M_{M} is the mass of the Moon.

At the surface of the Earth, $r = a$, and

$$V_M = -\frac{GM_M}{R}\left(1 + \frac{a^2}{R^2} - 2\frac{a}{R}\cos\chi\right)^{1/2}$$

$$= -\frac{GM_M}{R}\left\{1 + \frac{a}{R}\cos\chi - \frac{a^2}{R^2}\frac{1}{2}(3\cos^2\chi - 1)\ldots\right\} \qquad (4.4)$$

The first term of this series is a constant and produces no force upon the Earth. The acceleration of the Earth in the direction of a unit vector $\boldsymbol{n}$ due to the second term is

$$\frac{GM_M}{R^2}\frac{\partial\boldsymbol{n}}{\partial\boldsymbol{r}} = \frac{GM_M}{R^2}\cos(\boldsymbol{n},\boldsymbol{r})$$

But this is just the component in the direction $\boldsymbol{n}$ of the acceleration

$$GM_M/R^2$$

which is that due to the mutual gravitational attraction of the Earth and the Moon considered as ideal mass points and which keeps the Moon in its orbit about the Earth.

The third term in the potential,

$$\frac{GM_M}{R^3}\frac{a^2}{2}(3\cos^2\chi - 1)$$

represents the variation of the attraction of the Moon throughout the finite volume of the Earth and is known as the *tide-raising potential*.

The tidal term is a second-order spherical harmonic, for $\frac{1}{2}(3\cos^2\chi - 1) = P_2(\cos\chi)$ and it is convenient to write it as a product of harmonic functions of (Θ,Φ) and (θ,ϕ).

$P_2(\cos\chi)$ may be expressed as a series of harmonics in Θ, Φ and θ, ϕ, according to the formula (see Whittaker and Watson, 1940)

$$P_2(\cos\chi) = P_2(\cos\Theta)P_2(\cos\theta) + \sum_{m=1}^{2}P_2^m(\cos\Theta)P_2^m(\cos\theta)\cos m(\Phi - \phi)$$

$$(4.5)$$

Thus the tidal potential contains terms proportional to ϕ and 2ϕ and others independent of ϕ.

Now ϕ is the longitude of P measured not with respect to axes fixed in the Earth but with respect to axes fixed in space, and so it is equal to $\tilde{\omega}t + \lambda$, where $\tilde{\omega}$ is the angular spin velocity of the Earth. The part of the potential that varies at the speed $\tilde{\omega}$, namely

$$\frac{GM_M}{R^3}P_2^1(\cos\Theta)P_2^1(\cos\theta)\cos(\Phi - \tilde{\omega}t)$$

is called the *diurnal* term, while the part that varies at the speed $2\bar{\omega}$,

$$\frac{GM_M}{R^3} P_2^2(\cos\Theta) P_2^2(\cos\theta) \cos 2(\Phi - \hat{\omega}t)$$

is called the *semi-diurnal* term because it repeats after half a day. Because Φ, the longitude of the Moon, is not constant but is equal to nt, going through one revolution in about 28 d, the actual periods of the diurnal and semi-diurnal tides differ slightly from 12 and 24 h.

The coefficients of the diurnal and semi-diurnal terms are respectively

and
$$\left.\begin{array}{c} \dfrac{9}{4} \dfrac{GM_M a^2}{R^3} \sin 2\Theta \sin 2\theta \\[4mm] 9 \dfrac{GM_M a^2}{R^3} \sin^2\Theta \sin^2\theta \end{array}\right\} \qquad (4.6)$$

The orbit of the Moon is inclined at the angle i to the equator of the Earth and therefore Θ is given by

$$\cos\Theta = \sin i \sin \Phi$$

so that $P_2(\cos\Theta)$ undergoes a variation with the speed $2n$. The tidal component with this speed,

$$\frac{GM_M a^2}{R^3} P_2(\cos\Theta) P_2\cos\theta$$

is known as the *fortnightly tide*.

The Sun likewise gives contributions to the diurnal and semi-diurnal tides, namely

and
$$\left.\begin{array}{c} \dfrac{9}{4} \dfrac{GM_S}{R_S^3} \sin 2\theta \sin 2\Theta' \\[4mm] 9 \dfrac{GM_S}{R_S^3} \sin^2\theta \sin^2\Theta' \end{array}\right\} \qquad (4.7)$$

where M_S, R_S and Θ' are respectively the mass, radial distance and colatitude of the Sun.

Because n_S, the orbital speed of the Sun, is one-thirteenth that of the Moon, the periods of the solar diurnal and semi-diurnal tides differ slightly from those of the lunar tides.

The Sun, like the Moon, moves in an orbit inclined to the equator, producing a solar tide

$$\frac{GM_S}{R_S^3} P_2(\cos\Theta') \cos\theta$$

like the lunar fortnightly tide, but the period is half a year.

Since the gravitational potential V of the Earth at the surface is GM/a^2,

$$\frac{V_M}{V} = \frac{M_M}{M}\left(\frac{a}{R}\right)^3 \times \text{angular terms}$$

and

$$\frac{V_S}{V} = \frac{M_S}{M}\left(\frac{a}{R_S}\right)^3 \times \text{angular terms}$$

With

$$M = 6 \times 10^{24} \text{ kg}, \quad M_M = 7.4 \times 10^{21} \text{ kg} \quad \text{and} \quad M_S = 2 \times 10^{30} \text{ kg}$$

$$a = 6378 \text{ km}, \quad R = 3.8 \times 10^5 \text{ km} \quad \text{and} \quad R_S = 1.5 \times 10^8 \text{ km}$$

and

$$\frac{M_M}{M}\left(\frac{a}{R}\right)^3 = 6 \times 10^{-8}$$

$$\frac{M_S}{M}\left(\frac{a}{R_S}\right)^3 = 2.6 \times 10^{-9}$$

If the Earth were quite rigid, then the consequence of the tidal variation V_T in the potential would be that the height of the geoid above the solid surface of the Earth would simply vary by V_T/g, or aV_T/V but, because the Earth is not rigid, the solid surface distorts by hV_T/g and the potential changes by kV_T/g, so that the net change in the height of the geoid above the solid surface is

$$(1 + k - h)V_T/g$$

If the surface of the sea were to follow the geoid exactly, then the ocean tides would behave according to that rule. In fact, the surface of the seas differs greatly from expectation, partly because of the effects of currents and winds and partly because the currents corresponding to the tides themselves are greatly amplified in shallowing water, giving rise to tides in shallow waters and estuaries that are much greater than those in the deep oceans.

Since V_M as defined above is about $6 \times 10^{-8} V$, the amplitude of the equilibrium tide due to the Moon is 0.4 m, a value to be compared with tidal variations of some 10 m in certain estuaries.

Observations of tides in the oceans must be corrected for the attraction of the water mass itself and for Coriolis forces, but observations of diurnal and semi-diurnal tides in lakes should give reliable values of $1 + k - h$. The value found from Lake Tanganika is 0.56.

Another way of finding $1 + k - h$ is to measure the tilt of the land surface relative to the direction of gravity. The normal to the land surface has direction cosines

$$\frac{h}{V}\left(\frac{\partial V_{\mathrm{T}}}{\partial x}, \frac{\partial V_{\mathrm{T}}}{\partial y}, \frac{\partial V_{\mathrm{T}}}{\partial z}\right)$$

while those of the direction of gravity are

$$\frac{1+k}{V}\left(\frac{\partial V_{\mathrm{T}}}{\partial x}, \frac{\partial V_{\mathrm{T}}}{\partial y}, \frac{\partial V_{\mathrm{T}}}{\partial z}\right)$$

and the differential tilt therefore has direction cosines

$$\frac{1+k-h}{V}\left(\frac{\partial V_{\mathrm{T}}}{\partial x}, \frac{\partial V_{\mathrm{T}}}{\partial y}, \frac{\partial V_{\mathrm{T}}}{\partial z}\right)$$

The most common instrument for measuring tilts is the horizontal pendulum (figure 4.4). The design is very similar to that of the horizontal pendulum used as a seismometer but the sensitivity for the measurement of tidal tilts should be as

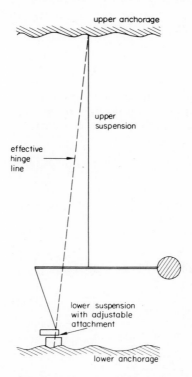

Figure 4.4 A horizontal pendulum

high as possible and the period as long as possible. It is advantageous to have a very long suspension for the pendulum, and the longest so far made has been set up in the Grotta Gigante, a limestone cavern some 100 m high near Trieste. Measurements of tilts are not in fact very satisfactory because the yielding of the ground is not governed just by the behaviour of the Earth as a whole but is very sensitive to local mechanical variations of the rocks from place to place while, in addition, the attraction to which a pendulum is subject includes that of nearby seas which may themselves undergo considerable tidal variations. Thus it is not surprising that tilt measurements give results that it is difficult to apply to the Earth as a whole and the detailed significance of which is often obscure.

The number k can be obtained separately from h in two ways. The change of potential kV must cause changes in the orbit of an artificial satellite and it has been possible to determine them for the daily tide (Kozai, 1967). Again, the redistribution of mass to which kV corresponds also causes a change in the moment of inertia of the Earth and therefore in the rate at which the Earth spins about its axis. The rate can be found very accurately by observing the positions of stars at times given by an atomic standard of frequency and it has been possible to detect a fortnightly variation (see section 4.5).

Consider now the variation in the vertical component of gravity due to the tidal potential. The potential of the Sun or the Moon, V_T, being of external origin, is proportional to

$$\frac{GM'r^2}{R^3} S_2$$

where S_2 is a second-order surface harmonic, and M' is M_M or M_S; on the other hand, the potential of the rearranged matter within the Earth, being of internal origin, is proportional to

$$kGM' \frac{a^5}{r^5} \frac{a^2}{R^3} S_2$$

at a radial distance r.

The total external potential may therefore be written as

$$V + V_T + k\frac{a^5}{r^5} V_T$$

The acceleration due to gravity is, to first order,

$$-\frac{\partial V}{\partial r} - \frac{\partial V_T}{\partial r} - k \frac{\partial}{\partial r}\left(\frac{a^5}{r^5} V_T\right), \quad \text{evaluated at } r = a$$

Now $\partial V_T/\partial r = 2V_T/r$ and so gravity is

$$g - \frac{2V_T}{r} - \frac{ka^5}{r^5}\frac{2V_T}{r} + 5k\frac{a^5}{r^6} V_T = g\left\{1 - \frac{2V_T}{rg}(1 - \tfrac{3}{2}k)\right\} \quad \text{at } r = a$$

But a changes to $a + hV_T$ because of the displacement of the surface and therefore the value of gravity changes to

$$\left\{ 1 - \frac{2V_T}{ag} (1 - \tfrac{3}{2}k + h) \right\}$$

as compared with the value $g(1 - 2V_T/ag)$ for an unyielding Earth.

Thus, by measuring the variations of gravity with the tidal periods, it should be possible to find $1 - \tfrac{3}{2}k + h$.

The variations are of order $2g(M_M/M)(a/R)^3$ for the lunar tide and $2g(M_S/M) \times (a/R_S)^3$ for the solar tide, or 0·1 mgal and 0·005 mgal respectively.

Modern gravity meters are amply sensitive enough to detect such variations although great care must be taken to insulate them from changes of temperature and pressure which, because they commonly have components that vary with a daily period, can give rise to apparent changes of gravity that could be confused

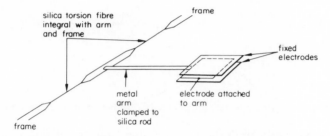

Figure 4.5 Gravity meter constructed by B. Block and R. D. Moore (1970)

with the real tidal changes. Recently gravity meters with high sensitivity and low noise have been developed especially for observations of tides and free oscillations (section 4.4). In one a very stiff torsional suspension is used (figure 4.5) so that the deflection is insensitive to mechanical noise and does not drift appreciably, and the necessary sensitivity of reading is obtained by using electronic circuits of low noise and high gain to display the change of capacitance between a fixed electrode and one attached to the suspended mass. A similar instrument has been constructed by R. V. Jones, In a second type of gravity meter (figure 4.6) a superconducting mass is supported by eddy currents in the magnetic field generated by supercurrents in a fixed coil. The whole apparatus is in liquid helium and the supercurrent, and hence the position of the suspended mass, is exceedingly stable. Again, the tidal changes of the acceleration due to gravity have been observed (Prothero and Goodkind, 1968). Variations of the acceleration due to gravity are less confused by local effects than are tilts but, as with the observation of tilts, changes of the acceleration due to gravity give the values of k and h that differ for different speeds of tide.

The number l can be found by direct observation with strain meters using quartz rods or laser interferometers such as are employed to detect seismic waves of very long period. Horizontal strain also affects measurements of the variation of latitude. Consider a place on the surface of the Earth from which the colatitude of the north pole (or the star Polaris) is observed relative to the direction of the local vertical. The latter will change in direction by

$$\frac{1+k}{V}\left(\frac{\partial V_T}{\partial x}, \frac{\partial V_T}{\partial y}, \frac{\partial V_T}{\partial z}\right)$$

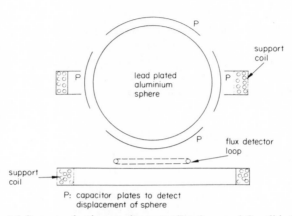

Figure 4.6 Superconducting gravity meter (Prothero and Goodkind, 1968)

while the angular position of the site measured from the direction of the north pole will change by $(l/V)(\partial V_T/\partial x)$ if x is in the meridian. Thus the apparent change in latitude will be

$$\frac{1}{V}(1+k-l)\frac{\partial V_T}{\partial x}$$

Regular measurements of the apparent latitudes of observatories forming an international network are made as part of the International Latitude Service (section 4.5) and, from analysis of the variations of latitude with tidal period, $1+k-l$ has been found.

Results of the measurements of the three Love numbers by different methods are summarized in table 4.1. The most reliable determinations are those of k

Table 4.1   Values of Love numbers

Quantity	Method	Value
k	orbits of artificial satellites	$0\cdot39 \pm 0\cdot05$
	change in length of day	$0\cdot34 \pm 0\cdot05$
	Chandler period	$0\cdot28 \pm 0\cdot02$
$1+k-h$	tides in lakes	$0\cdot56$
$1+h-\frac{3}{2}k$	gravity measurements	$1\cdot14$ to $1\cdot20$
$1+k-l$	variation of latitude	$1\cdot13$
l	Earth strain measurement	$0\cdot05$

from the fortnightly variations of the length of the day and from the variations in the potential detected by artificial satellites, while the least reliable are tilt meter determinations of $1 + k - h$. The following may be taken as representative values:

$$h = 0 \cdot 7$$
$$k = 0 \cdot 3$$
$$l = 0 \cdot 1$$

4.5 Free oscillations

Just as a bell is set into sustained oscillations when it is struck so, if the Earth is given an impulse by a large earthquake, it is set into free oscillations in which movements over the whole Earth are correlated, instead of, as with body waves or surface waves of high frequency, being almost independent at widely separated sites.

Ignoring the self-gravitational forces, the equations of motion are

$$\left. \begin{array}{l} \rho \ddot{u} = (\lambda + \mu) \dfrac{\partial \varDelta}{\partial x} + \mu \nabla^2 u \\[2ex] \rho \ddot{v} = (\lambda + \mu) \dfrac{\partial \varDelta}{\partial y} + \mu \nabla^2 v \\[2ex] \rho \ddot{w} = (\lambda + \mu) \dfrac{\partial \varDelta}{\partial z} + \mu \nabla^2 w \end{array} \right\} \qquad (4.7)$$

The equation of continuity must also be satisfied:

$$\dot{\rho} = -\rho \varDelta - u \operatorname{grad} \rho$$

Solutions in spherical polar coordinates are sought of the form $f(r) S_{mn}(\theta, \phi) e^{i\omega t}$ where S_{mn} is some surface harmonic. The solutions will have to satisfy the condition that forces vanish at the outer surface. The case of the uniform sphere, for which λ, μ and ρ are everywhere the same, was solved by Love (1911). For a given function $f(r) S_{mn}$, ω is determined by setting the surface forces to zero.

Because the properties of the Earth vary with radius, it is not possible to obtain an analytical solution and the equations must be integrated numerically.

Two fundamental types of oscillation can occur (figure 4.7). In one, part of the Earth twists relative to another. Such oscillations are called *toroidal* oscillations and involve no change of dilatation. Thus they show up as changes of strain or displacement at the surface of the Earth but, because there are no changes of density within the Earth, there are no changes of gravity at the surface. The other modes involved radial oscillations, changes of density and variations of gravity at the surface—such oscillations are called *spheroidal*. Any mode of oscillation will be characterized by the frequency and by the pattern of nodal surfaces—those surfaces on which the motion vanishes.

If n and m are respectively the order and degree of the surface harmonic part of the solution, and if the number of nodal spheres on which the radial function vanishes is l, the two classes of oscillation are labelled $_lS_n^m$ for the spheroidal modes and $_lT_n^m$ for the toroidal modes.

If it is assumed that the solutions are proportional to $e^{i\omega t}$ the equations of motion then read

$$
\left.
\begin{aligned}
-\rho\omega^2 u' &= (\lambda + \mu)\,\frac{\partial \Delta'}{\partial x} + \mu\nabla^2 u' \\[2mm]
-\rho\omega^2 v' &= (\lambda + \mu)\,\frac{\partial \Delta'}{\partial y} + \mu\nabla^2 v' \\[2mm]
-\rho\omega^2 w' &= (\lambda + \mu)\,\frac{\partial \Delta'}{\partial z} - \mu\nabla^2 w'
\end{aligned}
\right\}
\tag{4.8}
$$

and

$$
i\omega\rho' = -\rho\Delta' - u'\,\mathrm{grad}\,\rho
$$

where u', v', w' are u, v, w divided by $e^{i\omega t}$ and Δ' and ρ' are the corresponding quantities.

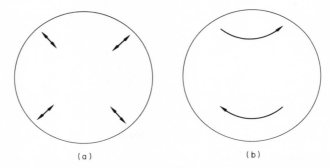

Figure 4.7 Mode patterns of free oscillations: (a) radial; (b) torsional

By writing $\rho = \rho_0 + \delta\rho$, where ρ_0 is independent of time but depends on the radius, and by assuming harmonic forms for $S_{mn}(\theta, \phi)$ the equations of motion reduce to equations for $f(r)$ which have solutions that satisfy the boundary conditions only if ω assumes a particular value. As stated before, the solutions have to be found numerically for realistic variations of λ, μ and ρ_0 with radius.

Because the values of ω depend on the particular variations chosen for λ, μ and ρ_0, a comparison of observed with calculated periods will provide a check upon a given model of the elastic properties of the Earth. It does not follow that different models may not give the same periods, at least within the accuracy of observation and calculation. In particular, the oscillations of high frequency have radial functions, $f(r)$, that are very small except in the outer layers of the Earth and so the periods of the high-frequency oscillations are insensitive to the properties of the deeper parts of the Earth.

So far the rotation of the Earth has been ignored, but the centrifugal body force should be included in the equations of motion. Alternatively, the equations may be transformed to a frame of reference rotating with the angular speed of the Earth. The situation is formally similar to that of the Zeeman effect in atomic spectra, where the forces exerted by an external magnetic field are equivalent to a transformation to a rotating frame of reference, and, just as the spectral lines of an atom are split into a number of components in those circumstances, so the modes of oscillation of the Earth are split. If v_0 is the speed of a mode for $m = 0$, the speed v_m, for the corresponding mode with the same n but non-zero m (which may range from $-n$ to $+n$) is

$$v_m = v_0 \pm m\tau\tilde{\omega}$$

$\tilde{\omega}$ is the spin angular velocity of the Earth and $\tau = 1/n(n + 1)$ for torsional oscillations, and has the values $\tau(2) = 0\cdot4$, $\tau(3) = 0\cdot2$, $\tau(4) = 0\cdot1$ for the first spheroidal modes.

Free oscillations have very long periods and it was only when seismometers of very long period or gravity meters of very high stability became available that it was first possible to detect them. A very large earthquake in Chile in 1960 produced records on quartz-rod strain gauges and on gravity meters being used for Earth-tide observations that gave the first values of free periods of the Earth. There are two powerful checks that the phenomena are correctly identified with free oscillations: first, no periods corresponding to toroidal oscillations are found in

Table 4.2   Periods of some free oscillations

(a) Spheroidal	
n	Period (min)
2	53·6
3	35·5
4	25·8
5	19·8
6	16·0
8	11·7
10	9·6
15	7·1
20	5·8
25	4·9
(b) Toroidal	
n	Period (min)
6	15·36
8	12·26
10	10·3
12	8·9
14	7·9
16	7·16
18	6·5
20	6·0

the gravity meter records, whereas periods of both types are found in strain and tilt records, as expected from the theory; and, secondly, the splitting of the modes arising from the rotation of the Earth has been found. A third check that can be applied with a growing number of observations is that the positions of nodes on the surface of the Earth should correspond to those expected theoretically. With recent very sensitive and stable gravity meters (section 4.3) free oscillations excited by earthquakes of seventh magnitude have been detected.

A typical set of periods is shown in table 4.2.

Periods of free oscillations of the Moon have been estimated in case they should be detected on a lunar seismometer.

4.6 Variations in the rotation of the Earth

The rotation of the Earth changes in two ways: the speed of rotation of the Earth about its axis and the direction of the axis of rotation both undergo continual changes. The variations are small and require very delicate observations for their detection but each type provides some information about the internal structure of the Earth.

The speed of rotation of the Earth is measured by comparison with the time scale established by an atomic standard of time. The unit of time, the second, is now defined by international agreement as the time taken for a certain number of cycles of the frequency (9129·6 MHz) corresponding to the difference of energy between two levels in the ground state of the caesium atom of mass 133—those in which the nuclear spin is parallel to and anti-parallel to the resultant atomic spin. The atomic definition of the second has replaced the previous astronomical one in terms of the time of rotation of the Earth upon its axis because comparisons of the rotation of the Earth upon its axis with the rotation of the planets (including the Earth) about the Sun showed that the rotation was not regular, and comparisons with time established at first with quartz crystal oscillators and later with atomic frequency standards have fully confirmed the variability of the speed of rotation of the Earth. In practice, a low-frequency signal is derived from the caesium frequency and is used to operate a clock against which telescopic observations of the times of passages of stars past the meridian are recorded. Most precise observations are made with *zenith tubes*, cameras which can see only a small area of the sky around the local zenith. The precision of the observations is limited entirely by the observations of the stars and particularly by the fact that it is rarely possible, because of atmospheric conditions, to obtain observations on every night.

The general behaviour of the rotation of the Earth is shown in figure 4.8. The dominant effects are variations with periods of one year and half a year, the former with an amplitude of 0·5 ms and the latter with an amplitude of 0·3 ms. The annual variation is caused by seasonal changes in the angular momentum of winds and currents, mainly winds, the momentum being greater in the southern hemisphere during the southern summer than it is in the northern hemisphere during the northern summer. Winds could make a small contribution to the semi-annual variation but are not the major cause. The major cause of the semi-annual variation in the length of the day seems to be tidal yielding of the solid Earth and consequent changes in the polar moment of inertia. The solar semi-annual tide is a zonal harmonic which corresponds to a variation in $C - A$ and hence in C (see Munk and MacDonald, 1960), and could account for about half the amplitude of the semi-annual variation of the length of the day. The change of C and the component of the variation of the length of the day are proportional to the number k, but k cannot be estimated from the data for the semi-annual term because of the additional meteorological contributions. But the lunar tide also has a zonal harmonic term, namely, the fortnightly term, which should give rise to a fortnightly variation in the length of the day unencumbered by meteorological effects.

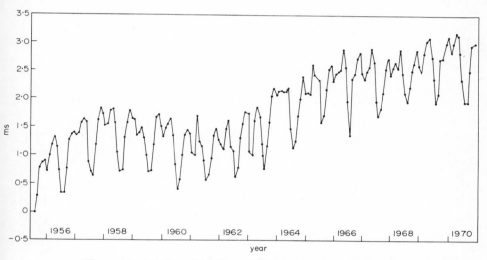

Figure 4.8 Variation of the length of the day (from Dr. L. Essen)

The fortnightly variation is indeed well observed and the value of k derived from it is given in table 4.1.

Changes in the direction of the pole of rotation of the Earth are obtained from astronomical observations of latitude, that is, of the angle between the local vertical and the axis of rotation. The fundamental method (figure 4.9) is to measure the distance between a star and the vertical, first when the star is closest to the vertical and then again twelve hours later when it is farthest from the vertical. From the two observations the latitude of the observatory and the angular distance of the star from the north pole can both be calculated, for the first measured angle is the difference of the colatitude and north polar distance, while the second is the sum. There are other special methods and instruments for the measurement of latitude regularly in use in observatories. In reducing the observations, allowance must be made for the refraction of the atmosphere and tidal changes in the local vertical among other effects and, in order to obtain uniform results, measurements are made as a matter of routine at certain stations, forming the International Latitude Service, that all lie on the 39th parallel (in the northern hemisphere) and that are equally spaced in longitude (figure 4.10). The uncertainty of a single observation of latitude is about 0·1 arcsec, and that of the mean of all observations in a month should be about 0·01 arcsec, but larger errors are found.

The direction of the axis of rotation as found from latitude observations shows small annual and semi-annual variations in the form of an elliptical path about the mean position and the annual wobble is to a first order accounted for by

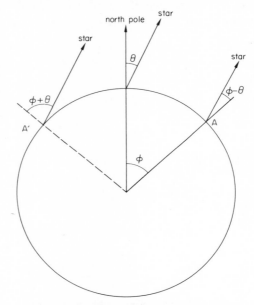

A, A′ positions of site separated by rotation of 180°

Figure 4.9 Principle of the measurement of latitude

changes in position of the atmosphere in the course of the year: air moves from the northern oceans in winter and piles up over Asia. The semi-annual wobble is not well determined and no cause is known for it.

After the seasonal variations in latitude are removed from the observations, it is found that there remains a wobble with a period of some fourteen months—the *Chandler wobble*, so-called after the discoverer, S. C. Chandler, a merchant of Cambridge (Mass.). The variations so far described in the length of the day and the direction of the axis of rotation bear a constant phase relation to the year or month, reflecting the constancy of the phase of the tide-raising and meteorological forces, but the Chandler wobble behaves differently, for after about ten years the variations are no longer in phase with each other, and the

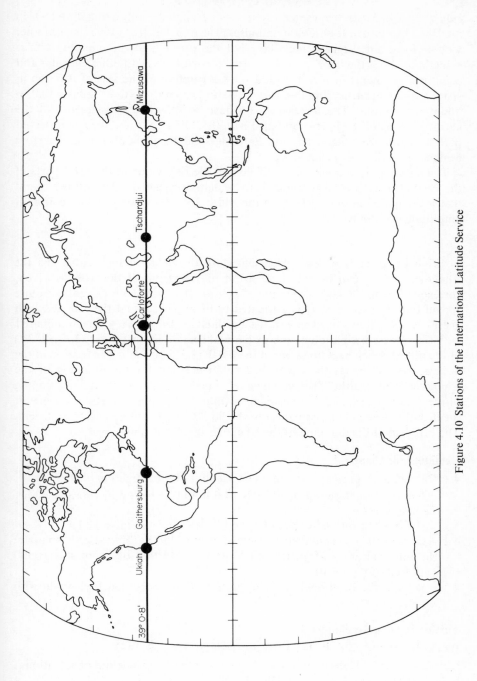

Figure 4.10 Stations of the International Latitude Service

behaviour is like that of a resonant system which is appreciably damped and which is excited by more or less random impulses. The impulses that excite the Chandler wobble have not yet been identified, but the period and the damping tell us something about the interior of the Earth. A rigid body with different polar and equatorial moments of inertia would show a rotation of the axis of rotation in space with an arbitrary amplitude and a period equal to $A/(C-A)$ days, that is, about ten months. The increase to fourteen months is a consequence of the elastic yielding of the Earth combined with the fluidity of the core. The observed period cannot be obtained if the Earth is solid throughout and is therefore evidence that the core is liquid.

The damping is a consequence of non-elastic behaviour in the solid parts of the Earth. If Q is the energy stored in an oscillating system divided by the amount that is dissipated in one cycle, then the width $2\Delta v$ of a resonance curve at half-amplitude is given by

$$2\Delta v = v_0/Q$$

v_0 being the peak frequency. The half-width of the resonance curve for the Chandler wobble can be calculated from the spectrum of the wobble which, corresponding to the lack of relation of phases after a delay of ten years, is quite spread out. The value of Q is estimated to be about 30, and the rate at which energy is dissipated somewhere in the Earth is 3×10^6 W, very small compared with the rate of dissipation of energy by tidal friction in the oceans (3×10^{12} W) or the rate at which heat flows out of the Earth ($2\cdot5 \times 10^{14}$ W). The friction of the oceans moving across the ocean bed could account for the damping of the Chandler wobble, although it would be very much less than the frictional damping of the oceanic tides, and non-elastic behaviour in the mantle might also account for it, but it seems that dissipation in the liquid core is unlikely to be sufficient. The problem of the damping of the Chandler wobble is quite open.

Examples for chapter 4

4.1 Calculate the precession of Mars on account of the attraction of the Sun.

The rotational period of Mars is 24·6 h and its period of rotation about the Sun is 687 d.

(Derive H from the values of J_2 and C/Ma^2 given in table 11.1.)

4.2 Assuming the Love numbers to be the same for Mars as for the Earth, compare the tidal variation of gravity on Mars (the Martian tides are solar only) with that of the Earth.

4.3 Calculate the rotational splitting pattern of the torsional free mode with $n = 3$.

Further reading for chapter 4

Block, B., and Moore, R. D., 1970, *J. Geophys. Res.*, **75**, 1493–505.

Kozai, Y., 1967, 'Determination of Love's number from satellite observations', *Phil. Trans. Roy. Soc., Ser. A*, **262**, 135–6.

Love, A. E. H., 1960, *Some Problems of Geodynamics* (Cambridge: Cambridge University Press).

Munk, W. H., and MacDonald, G. J. F., 1960, *The Rotation of the Earth* (Cambridge: Cambridge University Press), xix + 323 pp.

Plummer, H. C., 1960, *An Introductory Treatise on Dynamical Astronomy* (New York: Dover Publications), ix + 343 pp.

Prothero, W. A., Jr., and Goodkind, J. M., 1968, 'A superconducting gravimeter', *Rev. Sci. Instrum.*, **39**, 1257–62.

Whittaker, E. T., and Watson, G. N., 1940, *Modern Analysis*, 4th edn, (Cambridge: Cambridge University Press), 608 pp.

CHAPTER FIVE

The internal structure of the Earth

'What is your substance, whereof are you made?'
Shakespeare, *Sonnet* LIII

5.1 Introduction

Let us now try to put together the various ways of studying the Earth that have been explained in chapters 2, 3 and 4 and work out the mechanical properties of the Earth. Then let us see if we can understand these properties in the light of the behaviour of materials at the high pressures that exist in the interior, and let us try to infer the chemical composition of the different zones of the Earth. The data we have are the gravity field of the Earth, the velocities of seismic waves and the dynamical properties of the Earth as a whole, and from them it will be shown how the equations of state—that is the dependence of density upon temperature and pressure—may be derived for the materials in the different parts of the Earth. Having obtained the equations, they may then be compared with those of materials known in the laboratory; it is by no means a trivial matter to make the comparison because the temperatures and pressures readily accessible in the laboratory are far removed from those that are attained in the deep interior of the Earth, and it is only recently that experiments and theory have been able to close the gap in part.

All the phenomena discussed in the preceding three chapters depend in some way or another upon the variation of the density and elastic properties of the material of the Earth with radius. Most of the phenomena depend on some overall behaviour of a property and cannot be used to determine unambiguously the variation with radius. The mean acceleration due to gravity at the surface is an integral, $G \int \rho \, d\tau / D$ taken throughout the volume of the Earth; ρ is the density at some point in the Earth and D is the distance between that point and the point where gravity is being calculated. While many distributions of density may not give the observed value of the acceleration due to gravity, there is an infinite set of variations of density that will give the observed values. It is possible to be more explicit. The coefficients J_n, C_{nm}, S_{nm}, in the external potential are multipole moments of the density distribution;

$$J_n = \iiint r^n \rho P_n(\cos \theta) r^2 \, dr \, d\theta \, d\phi$$

Any aspect of a hypothetical distribution of density that can be expressed as multipole moments can be checked against the observed external potential but, if a hypothesis cannot be so expressed, then it cannot be checked against the potential. The multipole moments are weighted averages of the density throughout the Earth, the weighting factor being r^n for the $2n$-pole moment, and so the moments of higher order give greater weight to the densities in the outer parts of the Earth; it was similarly seen in the previous chapter that the periods of the higher free modes of oscillation are almost independent of the properties in the deeper parts of the Earth.

The harmonics in the gravitational potential and the periods of the free modes comprise sets of data that are in principle infinite in number, although only the lower members have so far been observed and those not always unambiguously.

There are a number of unique global properties that can be determined and that are likewise integrals of the properties throughout the volume of the Earth, namely, the mass and moment of inertia, the Love numbers and the period of the Chandler oscillation. They cannot be used to infer distributions of properties in any but the most general way, but they set conditions upon distributions that may be derived otherwise.

5.2 The core

It was seen in chapter 3 that the direct P waves are not observed at detectors lying between $105°$ and $140°$ from a source and that the reason is that at a radius of 3473 km the P-wave velocity decreases abruptly inwards. Now the coefficient J_2 in the harmonic expansion of the external gravity field is

$$\frac{C - A}{Ma^2} \quad \text{by McCullagh's theorem}$$

while the luni-solar precession of the Earth is proportional to the so-called dynamical ellipticity of the Earth, denoted by H and equal to $(C - A)/C$; if we take the ratio of these two quantities, we obtain the moment of inertia C in terms of the mass and radius of the Earth:

$$\frac{J_2}{H} = \frac{C}{Ma_2} \tag{5.1}$$

The numerical values are

$$J_2 = 1 \cdot 0827 \times 10^{-3}$$

$$H = 3 \cdot 275 \times 10^{-3}$$

leading to

$$C/Ma^2 = 0 \cdot 3306$$

This is a key result in the study of the internal structure of the Earth and the planets and it is important to understand how it can be obtained. The essence of the argument is this: the precession of the nodes of artificial satellites from which

J_2 is found and the luni-solar precession of the Earth from which H is found are both proportional to the *difference* of the polar and equatorial moments of inertia, C and A, and it is not possible from either by itself to obtain the separate moments of inertia. But the observed precession depends on the balance between that torque and the angular momentum of the satellite in the one case and of the Earth in the other; the former depends on the mass of the Earth and the latter upon the moment of inertia of the Earth and so by comparing the two effects it is possible to calculate the moment of inertia in terms of the mass of the Earth.

The ratio C/Ma^2 is important because it shows how strongly the mass of the Earth is concentrated towards the centre. The moment of inertia of a spherical shell (a sphere with all the matter concentrated at the surface) is $\frac{2}{3}Ma^2$, while that of a sphere of uniform density is $\frac{2}{5}Ma^2$ (example 5.1), but, if the mass is concentrated to the centre, the moment of inertia will tend to zero. The value of C/Ma^2 therefore gives an indication of how strongly the mass is concentrated towards the centre of the Earth. Table 5.1 gives some numerical values for a specific

Table 5.1   Values of C/Ma^2 for an Earth model

Ratio of Densities		C/Ma^2
2		0·367
3		0·340
	Earth	0·331
4		0·318

This model is divided into an inner and outer zone by a sphere of half the radius of the outer surface. The ratio of densities is the density of the inner zone divided by that of the outer zone.

model. It is supposed that a sphere is divided into two zones by a sphere with a radius equal to half the surface radius, thus corresponding to the division of the Earth into core and mantle. The outer shell of the model has a density of ρ_1 and the inner sphere has a density of ρ_2, and the table gives the values of C/Ma^2 corresponding to certain values of the ratio ρ_2/ρ_1, as well as the actual value found for the Earth. Evidently the density in the inner part of the Earth must be nearly three times as great as that in the outer parts. It will be shown in chapter 11 that corresponding results may be obtained for the Moon and for some of the planets.

The P-wave velocity α is equal to $\{(K + \frac{4}{3}\mu)/\rho\}^{1/2}$ and, since it falls by a factor of about 2 on crossing into the core, either ρ must increase or $K + \frac{4}{3}\mu$ must decrease, or perhaps both changes may take place. In fact, since the S-wave velocity is zero in the core, μ is zero in the core and then it is found that the density must also increase to account for the change of α. The boundary between the core and the mantle is therefore a surface at which both the seismic velocities and the density change sharply. The fact that μ is zero in the core suggests, but does not prove, that the core is liquid, in the sense that it cannot withstand shearing stresses, for it might be that μ could still be very small and the S waves in the core might have failed of detection. However, values of the Love number k as derived from the Chandler wobble, the fortnightly variation in the length of the day, and the fortnightly variations in the orbits of artificial satellites, show that the Earth must have a liquid core. Early calculations of k using elementary models and taking the rigidity to be that derived from seismological studies of the outer layers of the Earth gave results about 0·2, which are much too low. The discrepancies are not removed if the actual distribution of density obtained from seismic data is used, while keeping the rigidity in the inner parts about the same as that in the outer parts, and it is only if the rigidity of the core is less than 10^{-3} of the bulk modulus that the observed value of k can be obtained. No solids are known with such a small ratio of bulk modulus to rigidity and it is legitimate to speak of the core of the Earth as liquid. Some use has been made in the above arguments of data obtained from detailed seismic observations, but they are used only to refine the argument, which is that values of the moment of inertia, of the Love number k, and the seismic shadow effect combine to demonstrate that the Earth has a core, of radius about half the surface radius, composed of a liquid that is more than twice as dense as the solid outer layers.

5.3 The variation of mechanical properties with depth

It was shown in chapter 3 that the variation of seismic velocity with depth can be derived from the times of travel of P and S waves through the Earth and the general behaviour of the velocities α and β was given in figure 3.34. Now

$$\alpha^2 = (K + \tfrac{4}{3}\mu)/\rho$$

and

$$\beta^2 = \mu/\rho$$

so that

$$\alpha^2 - \tfrac{4}{3}\beta^2 = K/\rho \tag{5.2}$$

$\alpha^2 - \tfrac{4}{3}\beta^2$ is often denoted by Φ. K is defined by the relation

$$dp = \frac{K\,d\rho}{\rho} \tag{5.3}$$

and so

$$\frac{d\rho}{dp} = \frac{\rho}{\alpha^2 - \tfrac{4}{3}\beta^2} \tag{5.4}$$

This relation will now be converted to one relating density to depth instead of to pressure. Since the pressures are much greater than the strengths of rock materials, shearing stresses may be ignored; the interior is effectively in an hydrostatic state, for which

$$\frac{dp}{dr} = g\rho \qquad (5.5)$$

where g is the value of gravity at radius r, namely

$$g = \frac{GM_r}{r^2} \qquad (5.6)$$

Here M_r is the mass contained within the radius r. Accordingly

$$\frac{d\rho}{dr} = \frac{-GM_r\rho}{r^2(\alpha^2 - \frac{4}{3}\beta^2)} \qquad (5.7)$$

The basis of this equation (first derived by L. H. Adams and E. D. Williamson, 1923) is that the density varies only with pressure. It cannot, for example, be used if the chemical composition changes in some region and so its application to the study of the interior of the Earth is limited.

No account has been taken of the effect of thermal expansion. In fact, the value of K derived from the seismic velocities is the adiabatic incompressibility and the Adams–Williamson equation will be applicable provided the temperature gradient in the Earth does not depart too greatly from the adiabatic gradient.

The Adams–Williamson equation may be used to determine ρ and p starting from the surface and working inwards. In practice, the calculations start from the top of the mantle, it being supposed that the mass and other properties of the crust are well enough known from other procedures. At the outer boundary of the mantle, the enclosed mass, M_r, is known, being the total mass of the Earth less that of the crust. ρ is not known directly at the top of the upper mantle, but the isostatic balance of surface features and the densities of rocks thought to be derived from near the top of the mantle suggest that it is about 3320 kg/m³. At the top of the mantle, M_r is the mass of the Earth less the mass of the crust, and so can be calculated with good accuracy. r and $\alpha^2 - \frac{4}{3}\beta^2$ are also known and therefore $d\rho/dr$ may be calculated from the Adams–Williamson equation. The density at some small depth h is then

$$\rho - h\frac{d\rho}{dr} = \rho + h\frac{GM_r\rho}{r^2(\alpha - \frac{4}{3}\beta)^2}$$

and now $d\rho/dr$ can be calculated at the depth h, for ρ is known from the calculation just made and the mass M_r will be the mass below the top of the mantle less the mass of the shell of thickness h. In addition, the value of gravity at any depth follows from the formula

$$g_r = GM_r/r^2$$

and then the pressure may be found from the relation

$$p = \int_{a}^{r} g\rho \, dr$$

Finally, since α and β are known, λ and μ or K and μ may be calculated, and thus the density and all the elastic properties of the material are known as functions of either pressure or depth; in particular, the equation of state, that is the dependence of density upon the pressure at the temperature within the Earth, follows from the calculated values of density and pressure.

As with all numerical integration, errors in the data cause errors in the solution which increase with increasing distance from the starting point, and it is important to have some additional check upon the results. Again, were there no discontinuities, the observed moment of inertia ratio, C/Ma^2, would provide the check. The discontinuities within the Earth complicate the derivation of the density and make it less definite than it would be in their absence.

Suppose that the Adams–Williamson equation can be used down to the boundary of the core. Then at that point the total mass in the mantle will have been found and, by subtracting it from the total mass of the Earth, the mass of the core is found. Similarly, knowing the variation of density with radius, the moment of inertia of the mantle can be calculated and, subtracting it from the moment of inertia of the Earth, the moment of inertia of the core is found. Division gives the ratio, I/Ma^2, for the core. In his first application of this procedure, K. E. Bullen found (1936) that I/Ma^2 came out as 0·57, a value that is quite implausible, for it is much greater than the value for a core of uniform density and would imply that the core was almost a hollow shell. The way out of the difficulty is to suppose that in integrating inwards from the outside of the Earth some jump increase in density has been missed.

It was seen in chapter 3 that at about 400 km depth, where there is a rapid increase of α and $\partial\alpha/\partial r$, there is also a jump in density, by an amount initially unknown. Since the jump of density at the core boundary is unknown and since there is probably a jump of density at the boundary of the inner core, there are now too many unknowns for the problem of the density distribution to be solved using only the seismic velocity and the mass and moment of inertia of the Earth. The requirement that the moment of inertia ratio of the core should not exceed 0·4 sets a minimum value for the density of the core and a maximum value for the jump in density at 400 km. Given these densities and the corresponding masses within the radii at which they occur, the variation of density within each distinct zone of the Earth can be found by using the Adams–Williamson equation. Within the core μ is zero and

$$K = \alpha^2/\rho$$

The models of the density ρ and elastic moduli within the Earth based on the foregoing consideration are mostly due to K. E. Bullen (see Bullen, 1963). Bullen divides the Earth into seven zones as listed in table 5.2. Zones A to D comprise the mantle and zones E to G the core, G being the inner core. The zones F and G contain only a small proportion of the mass of the Earth and so assumptions about their densities can make little difference to the calculated densities throughout other zones. Bullen in his first calculations therefore took two extreme hypotheses; in one the central density had its lowest possible value of 12 300 kg m^{-3} and in the other the central density was 10 000 kg m^{-3} greater. He also supposed that, in the transition region C, the density followed a quadratic dependence on radius. Then he could calculate the densities to be those shown in table 5.3 (see also figure 5.1). The elastic moduli and the pressure for the

Table 5.2   Major zones of the Earth

Zone	Region (Bullen's notation)	Radius (km)	Characteristics	Density (kg m^{-3})	Material
crust	A	6370	complex structures	3300	rocks
		6340			
upper mantle	B C	5960 5370	changes of phase and composition	4700	iron–magnesium silicates
lower mantle	D' D"	3670 3470	little change of phase or composition	5700 9700	iron, aluminium, magnesium oxides
core	E F G	1390 1250	liquid transition solid	13000	predominantly iron and nickel

Table 5.3   Densities in Bullen's model A

Zone	Radius (km)	Density (kg m^{-3})
upper mantle	6270	3380
	6100	3550
	5900	3880
——	5370	4650
lower mantle	5000	4880
	4600	5100
	4200	5310
	3800	5510
		5660
——	3470	- - - - - - - - -
		9700
core	3400	9850
	2900	10500
	2400	11000
	1900	11600
——	1250	
inner core		12500

models are shown in figure 5.2. Bullen noticed that, if K and dK/dp are plotted against pressure they show an almost continuous variation (figure 5.3), despite the very great change in density at the core and other lesser discontinuities elsewhere, and he therefore postulated that they could in fact be taken to be strictly continuous. Adopting that postulate, he constructed a further set of models (Bullen, 1963); models of the first type are known as Bullen A models, and models using the compressibility–pressure hypothesis are known as Bullen B models.

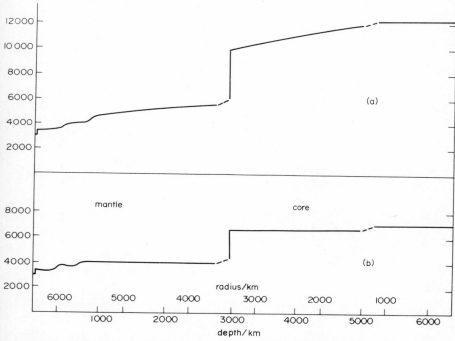

Figure 5.1 The variation of density within the Earth: (a) actual density; (b) reduced to zero pressure

The compressibility–pressure hypothesis by itself is insufficient to resolve all the ambiguities in the determination of the density from the seismic velocities and, in particular, the properties of the inner core may still vary within a wide range. Some progress towards resolving the uncertainties may be made by comparing the periods of the free vibrations of the Earth with those calculated on different models, but it is only the modes with the longer periods that provide much information since the amplitudes of the high-frequency modes are small in the inner parts of the Earth. The indications are that the properties of the upper part of the lower mantle are better represented by a model of type A rather than by one of type B, a result that implies that K cannot be quite continuous across the boundary of the core.

The Adams–Williamson equation was for some time the only means by which the density and elastic properties of the interior of the Earth could be studied but nowadays other methods are used. In the first place attention has increasingly been paid to evidence for changes in the composition of the material of the mantle and Bullen has emphasized the importance of a quantity that he calls η, as a measure of such changes.

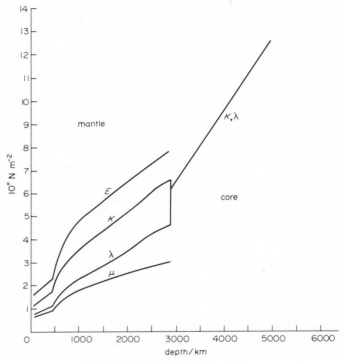

Figure 5.2 Elastic moduli within the Earth (Bullen model A)

By the definition of Φ,

$$K = \Phi\rho$$

and so

$$\frac{\mathrm{d}K}{\mathrm{d}p}\frac{\mathrm{d}p}{\mathrm{d}r} = \rho\frac{\mathrm{d}\Phi}{\mathrm{d}r} + \Phi\frac{\mathrm{d}\rho}{\mathrm{d}r} \qquad (5.8)$$

By the hydrostatic condition,

$$\mathrm{d}p = -g\rho\,\mathrm{d}r$$

or

$$\frac{\mathrm{d}p}{\mathrm{d}r} = -g\rho$$

and therefore

$$-g\rho \frac{\mathrm{d}K}{\mathrm{d}p} = \rho \frac{\mathrm{d}\Phi}{\mathrm{d}r} + \Phi \frac{\mathrm{d}\rho}{\mathrm{d}r}$$

or

$$\frac{\mathrm{d}\rho}{\mathrm{d}r} = -\frac{g\rho}{\Phi}\frac{\mathrm{d}K}{\mathrm{d}p} - \frac{\rho}{\Phi}\frac{\mathrm{d}\Phi}{\mathrm{d}r} \tag{5.9}$$

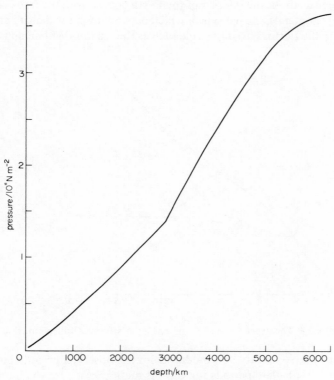

Figure 5.3 Pressure within the Earth

that is

$$\frac{\mathrm{d}\rho}{\mathrm{d}r} = -\frac{\eta g\rho}{\Phi} \tag{5.10}$$

where

$$\eta = \frac{\mathrm{d}K}{\mathrm{d}p} + \frac{1}{g}\frac{\mathrm{d}\Phi}{\mathrm{d}r} \tag{5.11}$$

If the material is of uniform composition, so that $K = \rho \mathrm{d}p/\mathrm{d}\rho$, then η is equal to 1, and if η takes some other value it means that K is not equal to $\rho \mathrm{d}p/\mathrm{d}\rho$, that is to say, that the density is changing for some other reason than just the effect of hydrostatic pressure. Now according to Bullen's results, dK/dp varies very little with material or with pressure, while throughout most of the Earth g varies only slightly. Thus η does not depend much on the model of density and elastic properties chosen for the Earth, but does depend directly on the observed quantity Φ. Bullen has therefore been able to examine the homogeneity of the Earth in some detail.

Other evidence about the interior of the Earth comes from the free oscillations of the Earth (chapter 4), the periods of which depend upon the density and elastic moduli within the Earth. No simple rule such as the Adams–Williamson equation

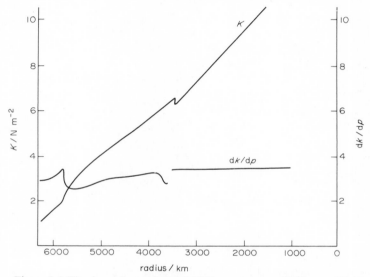

Figure 5.4 The dependence of K and $\partial K/\partial p$ on pressure (Bullen model A)

is available to derive density and moduli from the observed periods but instead it is necessary to calculate periods for various model Earths in which the density and moduli change in different ways with radius, and then to look for a set of calculated periods that match those observed. The same procedure can of course be used to match calculated travel–time curves with the observed ones. Prior to the development of digital computers, the very extensive calculations would have been impracticable but nowadays they are widely used, probably more so than the application of the Adams–Williamson equation.

Professor F. Press has used Monte Carlo methods in his studies. A model comprising values of α, β and ρ throughout the Earth is developed by a computer program, using random numbers to generate the actual values, and is then tested first against the travel times of P waves, then against the travel times of S waves, then against the mass and moment of inertia, and finally against periods of free modes, the model being rejected at any stage if it fails to satisfy a test. In this way models are chosen that satisfy the data and the results not only give estimates for the variation of properties within the Earth, but they also show within how wide a band the properties can lie at any depth and still satisfy the observed data. A more systematic inversion procedure has been developed in recent years by G. Backus, F. Gilbert and their collaborators. It is indeed important to emphasize that there is no one model that satisfies the data but there are many models consistent with the observations. Generally speaking, satisfactory models are similar to Bullen's models of type A but have the new feature, the consequence of better observations of S waves at moderate distances, that the seismic wave velocities go through a minimum at about 200 km.

Another approach to the problem of the interpretation of seismic results is to use laboratory data on properties of materials as directly as possible, in particular by looking for correspondences between values of Φ found from travel–time curves and values determined in the laboratory at high pressures and temperatures.

5.4 Materials and equations of state in the interior of the Earth (see Cook, 1972)

Having seen how the physical properties of the Earth may be estimated, let us now try to understand them in terms of the known or calculated behaviour of materials under very high pressures. At the surface of the Earth we see a bewildering variety of rocks made up of many different minerals and the physical properties also show great variation. The pressure is low and the properties of a solid are determined primarily by the forces between ions making up a crystal lattice, forces which determine the geometry of the lattice and therefore the density, elastic properties and the electrical behaviour of the crystal. The energies involved are of the order of 1 eV and are large compared with the product of pV, where p is atmospheric pressure (10^5 N m^{-2}) and V the atomic volume. 1 eV is $1 \cdot 6 \times 10^{-19}$ J, while the molecular volume of a crystal of molecular mass 140 (olivine) is $6 \cdot 6 \times 10^{-28}$ m^3 and thus pV is $6 \cdot 6 \times 10^{-23}$ J.

At pressures in excess of 10^9 N m^{-2}, the strain energy is greater than 1 eV so that the structure of the crystal lattice will have little effect on bulk properties, but the details of atomic structure are still significant, for they involve energies of 10 to 50 eV ($1 \cdot 6$–8×10^{-18} J). At pressures greater than 5×10^{10} N m^{-2} even atomic structure will have little effect on the bulk properties and the only significant factor is the total number of electrons associated with a nucleus—the atomic number.

As then we go in imagination deeper into the Earth, we expect to find properties determined first by details of the composition of rocks, then by the structures of crystal lattices and finally by mean atomic number. If the Earth were chemically homogeneous, a steady change in properties would ensue but the Earth is in fact divided into zones with distinct chemical composition (different mean atomic numbers) and its properties are determined by both pressure and composition.

The Earth is believed to consist almost entirely of compounds of silica, oxygen and metals, the mean composition possibly being similar to that of the chondritic meteorites thought to represent cosmic dust (table 5.4).

The materials that we see at the surface are mixtures of minerals, that is, of ionic crystals of silicon, oxygen and metals. Minerals occur in great variety and are found mixed together as a wide range of rocks, as has already been described in chapter 3.

The densities of silicate minerals increase with increasing metal content, as was shown in table 3.1 and correspondingly the densities of igneous and metamorphic rocks increase from the acidic to the ultrabasic. Seismic velocities increase in the same order (table 5.5).

Table 5.4   Composition of chondritic meteorites

Compound	Fraction in average chondrite
SiO_2	0·380
MgO	0·238
FeO	0·124
Fe	0·118
FeS	0·057
Al_2O_3	0·025
CaO	0·020
Ni	0·013
Na_2O	0·010
K_2O	0·002

The rocks of the upper part of the crust can be seen at the surface and the temperatures and pressures within the crust are not so different from those that can be attained in the laboratory that there is no great difficulty in identifying, at least in a general way, the materials that make up the crust. The continental crust has an average density and an average seismic velocity that correspond to a composition

Table 5.5   Densities and seismic velocities of rocks

	Density $(kg\ m^{-3})$	Compressional velocity $(km\ s^{-1})$
Granite	2667	5·0
Syenite	2757	5·5
Gabbro	2940	5·9
Dunite	3289	6·1

somewhere between acidic to intermediate, the upper part being more nearly acidic and the lower part more nearly intermediate if they can be separated. The crust below the oceans is denser and the seismic velocities are greater and the composition must be more nearly that of basalt. The structure of the crust is discussed in chapter 10.

When we attempt to identify the material of the mantle, we come up against the difficulty that the pressures and temperatures throughout much of the mantle are far greater than those that can be attained in an ordinary laboratory and also that, while we can see specimens of rocks that are very probably similar to those of the upper part of the crust, even the top of the mantle is inaccessible to us so that we do not have a starting point as we do for the crust. However, we do know the density of the uppermost part of the mantle from the conditions for isostatic balance, namely 3300 kg m^{-3}, and we know that that value falls in the range of the density of olivine. Further evidence comes from the consideration that most igneous rocks are derived from the uppermost layers of the mantle by a succession of steps of melting and fractional crystallization, and much is now known from laboratory investigations of the course followed by such steps, so that it is possible to get a good idea of the composition of the presumed original mantle material from the igneous rocks found in the crust. A particular mixture of olivine and pyroxene, called pyrolite, seems a likely composition for the upper part of the mantle.

Methods are now available in specialized laboratories for the study of solids at pressures similar to those attained in the mantle. By immersing the material in a suitable liquid, hydrostatic pressures can be applied, as was done by Bridgman in his pioneer studies. Higher pressures can be attained if the solid is compressed between anvils, although the pressure is not then hydrostatic but is of the type called uniaxial. Hydrostatic pressures up to $2 \cdot 5 \times 10^9$ N m^{-2} and uniaxial pressures up to 2×10^{10} N m^{-2} can be attained; they correspond to depths of 60 and 500 km respectively. High temperatures can also be produced at the same time as the high pressure. Equipment is now available in which X-ray diffraction measurements may be made upon a sample under high pressure so that the change of density with pressure may be found from the change in lattice spacing of the crystal, while the elastic moduli may be found from measurements of the speed of ultrasonic waves in the specimen. The variation of density, seismic velocity, the parameter Φ and the ratio K/μ as the pressure is changed are thus now available for various minerals that may occur in the mantle.

Quite early in his work on high pressures, Bridgman found that some crystals changed their structure under a sufficiently high pressure and went over to a more dense form. Quite a number of such phase transitions have subsequently been found and, in particular, it is known that olivine can change from the hexagonal crystal structure in which it normally occurs to a high-pressure form having a cubic structure.

Suppose a pressure p is applied to a crystal. The mechanical work done will be $p\ \delta V$ where δV is the change of volume, that is

$$\frac{V_0}{K} \times \frac{1}{2}\ p^2$$

where V_0 is the volume at zero pressure and K is the bulk modulus.

At the same time, the internal energy increases because the ions come closer together.

Now suppose that the crystal can change to a different form in which the ions are closer, therefore the internal energy is greater. The new form will not exist at zero pressure but, because the difference in work done by the external pressure is proportional to the difference of volume multiplied by the square of the pressure, that is, to $\Delta V_0 p^2/2K$, the denser form of crystal must be the more stable one at a sufficiently high pressure.

When the 20° discontinuity was first discovered it was suggested by J. D. Bernal and H. Jeffreys (see Jeffreys, 1970, p. 195) that the increase of density at a depth of 400 to 500 km might correspond to a change of olivine from the hexagonal to the cubic form. At the time the suggestion was made, the cubic form was not known but the cubic form of the germanium analogue of forsterite, Mg_2GeO_4, had been prepared, although not in large enough crystals for the density to be determined. Subsequent work, especially by A. E. Ringwood on crystals similar to olivine, has culminated in the discovery of the high-pressure form of olivine in the Tenham meteorite (Binns et al., 1969). The difference between the densities of the hexagonal and cubic (or spinel) forms of olivine is 350 kg m^{-3} while the change of density at the 20° discontinuity derived from seismic data is 700 kg m^{-3}.

The general conclusion from comparisons between laboratory and seismic data is that the mantle is composed of a material having the composition somewhat like olivine and that the change of density at 400 to 500 km is due to a change to a high-pressure form. Detailed studies show that the mantle is not of uniform chemical composition, the ratio of iron to magnesium increasing with depth. In addition, the effect of the increase of temperature with depth cannot be ignored and is probably responsible for the decrease in the shear wave velocity at about 200 km that has been discovered as results of studies of surface waves.

As the pressure increases towards that at the bottom of the mantle and within the core, it ceases to be possible to make experiments in which such pressures are maintained in the laboratory. Furthermore the dependence of the properties of substances upon pressures undergo a change, for at low pressures the crystal structure of a mineral has a large effect and the properties of minerals of similar composition but different structure may differ significantly. At high pressures, however, the properties are increasingly determined by the atomic volumes or mean atomic numbers of the constituents. Thus new methods of study are required and the results must be discussed in slightly different terms from those used for the upper mantle.

Pressures up to 2×10^{11} N m^{-2} can be established for times of the order of a few microseconds in samples exposed to an explosion. If an explosive attached to one face of a slab of mineral or metal is detonated, a wave of high pressure passes through the slab and, from measurements upon it, an equation of state for the high pressure can be obtained for the material of the slab. A diagram of the experimental arrangement is shown in figure 5.5. Best results are obtained by having the explosive drive a metal plate against the face of the sample rather than by attaching the explosive itself directly to the sample, and the plate arrangement is shown in the diagram. It is important that the impulse from the explosion should be applied simultaneously over the face of the sample and the charge is constructed to ensure that.

If the pressure applied to the slab by the impact of the moving plate is sufficiently great, the material of the slab will no longer deform elastically but a shock wave will be propagated through the slab at a velocity in excess of the speed of elastic waves. Let the pressure before the shock arrives be p_0 and let the density be ρ_0. After the shock has passed, let the pressure be p_1, let the density be ρ_1, and let the material be moving with a velocity u_1. Let the shock wave move with a velocity u_s.

The speed of the shock wave is obtained from the conditions that mass, momentum and energy are conserved in the passage of the shock.

The rate at which mass enters the shock is $\rho_0 u_s$ per unit area, and the rate at which it leaves is $\rho_1(u_s - u_1)$. The conservation of mass therefore tells us that

$$\rho_0 u_s = \rho_1(u_s - u_1) \qquad (5.12)$$

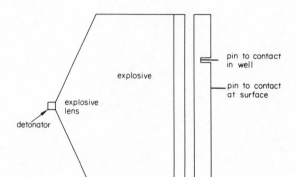

Figure 5.5 Arrangements for shock wave experiments. The velocities of the material and of the shock wave are found by timing electrically the instants of contact between the sample and the pins

The rate at which momentum changes in the passage of the shock is equal to the net change of velocity u_1, multiplied by the rate of passage of mass into the shock, namely, $\rho_0 \times u_s$, and this rate of change of momentum must equal the difference of pressure across the shock. Accordingly

$$p_1 - p_0 = \rho_0 u_s u_1 \qquad (5.13)$$

Lastly, the power supplied per unit area of the shock front must be equal to the rate at which the energy of the material in the front is changing. The power is $p_1 u_1$ and the rate of change of energy is equal to the rate at which kinetic energy is gained, that is, $\frac{1}{2}u_1^2 \times \rho_0 u_s$ plus the rate at which the internal energy of the material changes. If the internal energies per unit mass are E_1 and E_0, then the rate of change of internal energy is $\rho_0 u_s(E_1 - E_0)$ and so

$$p_1 u_1 = \frac{1}{2}u_1^2 \times \rho_0 u_s + \rho_0 u_s(E_1 - E_0) \qquad (5.14)$$

It is convenient to use the reciprocal of the density, namely the specific volume V.

Equation (5.12) may then be rewritten as

$$\frac{V_1}{V_0} = \frac{u_s - u_1}{u_s}$$

or

$$\frac{u_1}{u_s} = \frac{V_0 - V_1}{V_0} \tag{5.15}$$

Then, using equation (5.13),

$$p_1 - p_0 = \frac{u_s u_1}{V_0} = u_s^2 \frac{V_1 - V_0}{V_0^2}$$

or

$$u_s = V_0 \left(\frac{p_1 - p_0}{V_1 - V_0} \right)^{1/2} \tag{5.16a}$$

and

$$u_1 = (p_1 - p_0)(V_1 - V_0)^{1/2} \tag{5.16b}$$

On substituting these velocities in the energy equation (5.14), the difference of internal energy is found:

$$E_1 - E_0 = \tfrac{1}{2}(p_1 + p_0)(V_1 - V_0) \tag{5.17}$$

This result is known as the *Rankine–Hugoniot* equation and gives the difference of internal energy in terms of the values of pressure and specific volume along the shock front. The relation between p and V is not one at constant internal energy since E_1 and E_0 are different and so the relation is not an adiabatic one. Nor is it isothermal, but it is possible to reduce it to a pressure–volume ratio at absolute zero if the ratio γ, Gruneisen's ratio, is known for the material. γ is equal to

$$\alpha K / \rho C_v$$

where C_v is the specific heat at constant volume, K is the bulk modulus, α is the coefficient of thermal expansion and ρ is the density. It may not always be possible to calculate γ for the shock conditions from data at room temperature or ordinary pressures, but it is sometimes possible to derive it from shock wave measurements themselves.

The foregoing equations enable the change of pressure and the change of specific volume along the Hugoniot characteristic to be found from measurements of the velocities u_s and u_1, for

$$\frac{V_1}{V_0} = \frac{u_s - u_1}{u_s}$$

and

$$p_1 - p_0 = \frac{u_s u_1}{V_0}$$

The essential measurements thus consist in the determination of the velocity of the shock, u_s, and the velocity of the material behind the shock u_1. The former is obtained from the time of the arrival of the first pressure pulse through the sample of material, an event that may be detected by electrical contact between the face of the sample and a small adjacent electrode or, optically, by the light emitted by gas when it is compressed between the face of the sample and a window close to it. By arranging for contacts or windows of different depths in the sample it is possible, as shown in figure 5.5, to measure both u_s and u_1.

Many experiments on a range of metals have been performed in recent years in the U.S.A. and in Russia, the maximum pressures attained of the order of 5×10^{11} N m^{-2}, exceeding those at the centre of the Earth. Thus experimental data are available for materials likely to be similar to those in the core, not only for the variation of density with pressure but also for the variation of the bulk modulus and of the velocity of sound.

Figures 5.6, 5.7 and 5.8 show some typical results for density, bulk modulus and compressional velocity, the important characteristic of which is that at pressures of the order of 10^{11} N m^{-2} many elements show a generally similar change of properties with pressure, but with a systematic dependence upon atomic number. Pure iron has rather too high an atomic number to reproduce the behaviour of the core. If about 10 % of nickel is mixed with the iron (corresponding to cosmic abundances) the discrepancy is worse, and it must be supposed that some lighter material is also present. Silicon and sulphur have been suggested.

In the core the pressure is so high that the only significant parameter is mean atomic number and so it is possible from the shock wave data to make a rather confident estimate of the chemical composition.

Some idea of the equation of state of a substance at very high pressure can be derived theoretically. It has recently been seen that, when the pressure on certain crystals is high enough, the form of the crystal changes to one that occupies less volume. At yet higher pressure, a further reduction in the spacing of the ions can be achieved if electrons are detached from the ions and form a free electron gas. The crystal is then a metal. While the distinction between metals and non-metals is usually quite sharp under ordinary laboratory conditions, some elements and compounds are known in metallic and non-metallic forms, for example grey and

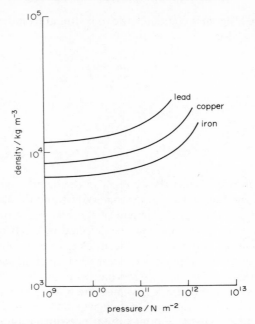

Figure 5.6 Shock wave results—density as a function of pressure

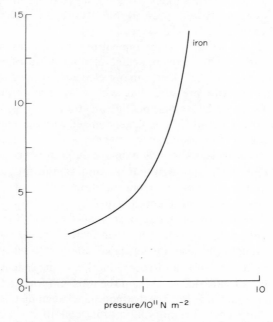

Figure 5.7 Shock wave results—bulk modulus as a function of pressure

white tin. It was suggested some years ago by W. H. Ramsey that the core of the Earth might be a metallic form of olivine but, as knowledge of the equation of state of iron has been obtained experimentally, and as knowledge of the properties of non-metal to metal transitions has grown (see Mott, 1968, for a review), Ramsey's idea seems less probable than the alternative, that the core is composed mainly of iron, and that the transition from the mantle to the core is one of composition and not of structure. The hypothesis of the transition to a metal under high pressure does, however, seem to be applicable to the planets Jupiter and Saturn (chapter 11).

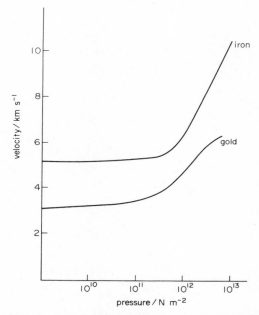

Figure 5.8 Shock wave results—compressional wave velocity as a function of pressure

At pressures somewhat greater than core pressures, spin–orbit interactions of electrons in an atom are insignificant compared with the total energy and it is possible to replace the actual set of electrons in an atom, with their complex structure, by a structureless cloud of electrons occupying the same volume as the actual atoms. The pressure of the electrons in a sphere of that volume is calculated from Poisson's equations, as first suggested by L. H. Thomas and E. Fermi. All the electrons of an atom are supposed to be associated with the nucleus, which is not quite true for a metal since some of the electrons will form a free Fermi gas. In the circumstances of the Thomas–Fermi model, molecules cannot form and if more than one element is present it is sufficient to use the mean atomic number and mean atomic volume.

Electron velocities are supposed to be non-relativistic but, in an extension to the original Thomas–Fermi model, the electrons may be taken to obey Dirac's equation by including exchange forces.

The result of the theory is that the pressure is given by

$$p = \frac{h^2}{5m} \left(\frac{3}{8\pi}\right)^{2/3} \frac{Z^{5/3}}{V^{5/3}} \left\{1 - \frac{2\pi me^2}{h^2} (4ZV)^{1/3} \dots\right\}$$

which can be seen to depend only on atomic number Z and atomic volume V.

Pressures in the core are too low for the Thomas–Fermi–Dirac model to be

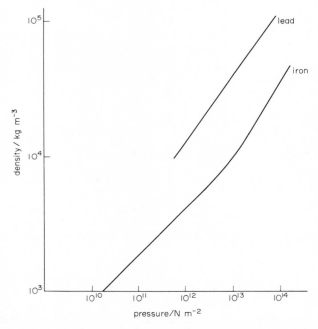

Figure 5.9 Curves of density as a function of pressure for Thomas–Fermi–Dirac models

strictly applicable but it is possible to join up equations of state from shock wave studies with those from the model in a plausible way as shown in figure 5.9. The main difference between the experimental and theoretical curves is that the latter have a much greater slope which would seem to start at pressures just greater than that at the centre of the Earth.

Recent seismic results and studies of the overtones of free oscillations have shown that the value of the shear wave velocity, β, in the inner core is about 3 km s^{-1}, showing that μ is finite. The inner core is thus definitely solid, although relatively softer than lead or gold under room conditions.

It may at first sight seem that, if the inner core is solid, then its composition must differ from that of the outer core because its temperature is higher, so that, if the outer core is liquid, then an inner core of the same composition would also be liquid. The argument neglects the effect of pressure upon the melting temperature. The melting temperature of metals rises appreciably with rise in pressure and F. E. Simon showed that many of the data followed a semi-empirical rule.

$$\frac{p}{a} = \left(\frac{T_p}{T_0}\right)^c - 1$$

where T_0 is the melting temperature at zero pressure, T_p that at the pressure p and a and c are constants that must be determined experimentally for each different material. Another semi-empirical law (Kraut and Kennedy, 1966) states that the melting temperature is a linear function of the specific volume, and if the specific volume at core pressures is obtained from the shock wave equation of state of iron then the melting temperature at the boundary of the inner core is found to be about 3700 °K. The essential point is that this is a temperature quite likely to obtain at that depth in the Earth, and that, if pressure causes the melting temperature to rise more rapidly than does the actual pressure, then the inner core would be solid.

An important result of the study of the equation of state of the material in the interior of the Earth is that the bulk modulus is a smooth and very nearly linear function of pressure. Bullen used the result to construct a number of valuable models of the density within the Earth but, with more information from seismology and from the study of the free modes of oscillation of the Earth, it is no longer necessary to rely upon the hypothesis in working out the density within the Earth and indeed it can now be seen that within the Earth there are small departures from any simple variation of bulk modulus with pressure. Nonetheless the linear variation of bulk modulus with pressure is a very good approximation and if it may be extended to the other planets may be used to investigate their properties in the absence of such detailed seismic data as we possess for the Earth, In fact it seems to depend on particular circumstances in the Earth (Cook, 1972) and its extension to other planets must be made cautiously.

5.5 Angular variations in the properties of the Earth

The model of the Earth developed in the preceding sections by combining geophysical observations with experimental and theoretical studies of properties of material at high pressures is based among others upon the hypothesis that the Earth is in a hydrostatic state and that shear stresses are negligible. While shear stresses are undoubtedly much less than normal pressures throughout most of the Earth and while, compared with other uncertainties, the neglect of shear stresses is for this purpose quite unimportant, it is equally clear that shear stresses do exist in the outer parts of the Earth. Shear stresses arise when the properties of the Earth vary with angular position as well as with radius, and in this section some of the evidence for the magnitude of shear stresses is considered and the strength required in the Earth is discussed in the light of experimental and theoretical knowledge of solids.

That the Earth is not in a hydrostatic state appears from the very fact that land appears above the surface of the water, for if the Earth were subject to no shear stresses then surfaces of constant geopotential would also be surfaces on which the density was constant. In addition, the density would always increase as the radius decreased and so the Earth would consist of a sequence of spheroidal layers each of constant density and, in particular, the waters would cover the land to a uniform depth. The fact that we exist as land animals shows that the material of the Earth has the strength to support non-hydrostatic stresses.

The actual value of the polar flattening f is also evidence that there are shearing stresses in the Earth for, if the Earth were in the hydrostatic state and surfaces of constant potential were also surfaces of constant density, then it would be possible to calculate the flattening from the rate of spin and the polar moment of inertia according to a theory given by Sir George Darwin (a son of Charles Darwin). Darwin's result is

$$\frac{m}{f} = \frac{2}{5} + \frac{5}{2}\left(1 - \frac{3}{2}\frac{C}{Ma^2}\right)^2 \tag{5.18}$$

where the quantity m was defined in chapter 2.

The calculation that gives C/Ma^2 as the ratio of the observed quantities J_2 and H does not depend on any assumption about the hydrostatic state of the Earth, and so we may substitute that value in Darwin's formula to find the value of f to be expected for an Earth in the hydrostatic state. The value so found is somewhat less than $1/299$ (see example 5.2) whereas the actual value is $1/298\cdot25$.

On the hydrostatic theory, J_4 would be of the order of J_2^2, J_6 would be of the order of J_2^3 and all other terms in the spherical harmonic expansion of the gravity potential would be zero. The fact that they are not shows that there are irregularities of density within the Earth that would not be present were it in the hydrostatic state. The best way of showing the results is as a map of the elevation of the geoid above the spheroid with the flattening expected on the hydrostatic assumption, for that will include the J_2 term as well as all the others. Such a map is shown in figure 5.10.

Other information about departures from the hydrostatic state comes from seismology. A great deal is known of the structure of the crust and how it varies from place to place and especially as between continents and oceans (see chapters 3 and 10) but so far not very much is known about lateral variations within the mantle although both travel times of body waves and dispersion curves of surface waves indicate that there are variations in velocity from place to place. In particular, low velocities in the mantle appear to lie beneath the mid-oceanic ridges (chapter 10). The low-velocity layer in the mantle may also be evidence of shearing stresses for if it is, as some think a layer of low density as well, it would have to be supported by shear stresses.

Given a load upon the surface of sphere or at some specified depth within it, it is possible to work out from the equations of elastic equilibrium the stresses

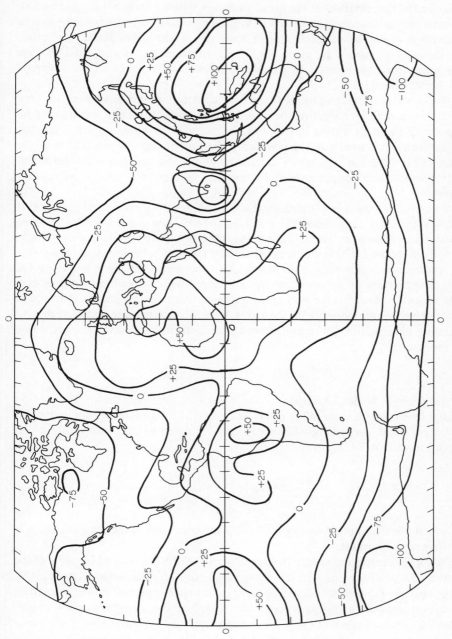

Figure 5.10 Map of N, the elevation of the geoid, above a spheroid of flattening 1/299

that are set up. Unfortunately the depth of the density variations that are responsible for the deformation of the geoid shown in figure 5.10 is not known and as pointed out in chapter 2, the density variations could be put anywhere over a wide range of depths, certainly almost down to the core. Thus a unique solution for the stresses cannot be given but, if the strain energy is supposed to be a minimum, then the stresses can be determined and would be of the order of 10^7 N m^{-2} (Kaula, 1968).

Where the non-hydrostatic loads are thought to be supported depends very much on how the strength of the mantle decreases with depth. Loads may be supported statically by the strength of the material, or they may be supported dynamically by currents in the mantle; in the former case the strength must be high, in the latter case the material must be weak. The strength of a solid varies with pressure and temperature and some idea of that behaviour must now be given.

When solids are subject to stresses that are sufficiently great or that are applied for a sufficiently long time, the resulting deformation does not vanish when the stress is removed as it does for an elastic deformation, but there is some permanent displacement. There are two well-defined types of behaviour that, as a result of studies over many years, are well understood for metals, one occurring for fast deformations at temperatures low compared with the melting temperature, and the other occurring for slow deformation at temperatures greater than about half the melting temperature. In the former case, work hardening occurs, the metal resisting deformation more and more as the deformation increases, according to the law

$$\varepsilon = t^n$$

first given by E. N. daC. Andrade. ε is the strain, t the time from the first application of the stress, and n is an index equal to about $\frac{1}{3}$. In high-temperature creep, on the other hand, the rate of change of strain is related to the temperature by an equation such as Eyring's:

$$\frac{d\varepsilon}{dt} = 2A \exp\left(-\frac{U_0}{kT}\right) \sinh \frac{\beta\sigma}{kT}$$

in which σ is the applied stress, A, U and β are constants of the material and k is Boltzmann's constant.

Work hardening and steady state creep can both be understood in terms of the movement of dislocations in the crystal structure of a metal (see Kittel, 1968). The ideal crystal consists of a perfectly regular array of atoms or ions in a lattice structure but such a crystal never occurs in nature and all crystals show imperfections in which some atoms are missing so that the crystals consist of regions in which the lattice structure is perfect, separated by lines across which atoms are displaced relative to the ideal structure. Such lines are called dislocations (figure 5.11). They do not extend right throughout the volume of a crystal but end on other dislocations or at alien atoms if the material is not chemically pure. When a stress is applied to a crystal, the dislocations move through the material and as a result the crystal is deformed. The elementary step of movement of a dislocation is through one lattice spacing and the rate of deformation is therefore given by the rate at which dislocations move, that is, it is equal to the number of dislocations in the crystal multiplied by the chance of a dislocation moving in unit time. Now the movement of a dislocation requires the supply of energy U

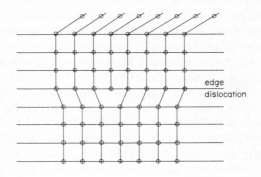

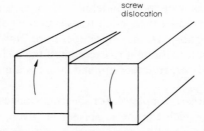

Figure 5.11 Dislocations in a crystal

per atom and the chance of that energy being supplied by the thermal motions of the atoms is

$$\exp(-U/kT)$$

In the absence of an applied stress, a dislocation may move randomly in any direction and there is no net deformation of the crystal but, if a stress is applied, the energy to move a dislocation in the direction of the stress is reduced, to a first approximation in a linear manner:

$$U = U_0 - \beta\sigma$$

where σ is the stress and β and U_0 are constants.

Eyring's equation follows from these assumptions.

The material of the Earth is certainly not composed of single crystals and so, although experiments in the laboratory show that rocks do creep according to the Eyring equation, it is unlikely that the movement is controlled by dislocations within crystals. However, it is found that polycrystalline metals also follow the Eyring equation, movement occurring through the displacements of the boundaries between grains. Just as for dislocations within crystals, a definite energy must be supplied to move the boundary between crystals, and so creep through the movement of boundaries depends on temperature and stress in the same way as does creep through the movement of dislocations within crystals.

The movement of a dislocation in general creates other dislocations and the greater the number of dislocations in a crystal, the greater the energy required to move a dislocation because dislocations tend to be fixed where they end on other dislocations (pinning). It would be expected therefore that, as the number of dislocations grows with the increasing deformation of the crystal, there will come a time when the applied stress is no longer adequate to deform the crystal further were it not for atomic diffusion. The material is then said to be work-hardened. The Andrade equation describes the approach to this condition. Polycrystalline materials similarly show work hardening and no doubt rocks in the crust of the Earth behave in that way, giving rise to a finite strength at low temperatures.

At higher temperatures, however, diffusion of atoms within a crystal or between crystals destroys dislocations and grain boundaries and, if the rate of destruction is high enough, dislocations will move indefinitely under an applied stress. Detectable steady state creep under these conditions is usually observed at temperatures greater than half the absolute melting temperature. Now the temperature throughout most of the mantle is thought to exceed half the melting temperature (see chapter 7) so that steady creep under applied stress is almost certainly important geophysically.

How, in those circumstances, is it possible for shearing stresses to be supported in the Earth? The answer, which cannot at present be given, must depend on where the shear stresses occur and on the variation of temperature within the Earth. If all the shearing stresses were confined to the crust then it might be that they would be supported by the finite strength of work-hardened material, but the major irregularities of gravity are not correlated with the distribution of continents and oceans and it seems likely that the corresponding shearing stresses are developed within the upper mantle where the behaviour is probably governed by steady state creep. The material may nonetheless have an effective strength over a sufficiently long time because the rate of change of strain for the stresses in question may be so low as to have no significant effect in that time. Suppose that the least rate of strain that can be detected by a study of gravity anomalies or otherwise is $\dot{\varepsilon}_0$. Then any stress less than

$$\sigma_0 = \frac{kT}{\beta} \operatorname{arcsinh}\left\{2A\dot{\varepsilon}_0 \exp\left(\frac{U_0}{kT}\right)\right\} \tag{5.19}$$

will produce no detectable strain and that stress may be regarded as the strength of the material. σ_0 decreases with temperature so that, if the temperature in the mantle is high enough, the effective strength would be too low to support the loads represented by the gravity anomalies.

If, on the other hand, the temperature is sufficiently close to the melting point, the Eyring equation becomes approximately

$$\dot{\varepsilon} = 2A \exp\left(-\frac{U_0}{kT}\right)\frac{\beta\sigma}{kT} \tag{5.20}$$

that is to say the material behaves as a liquid with a viscosity equal to

$$\frac{2A\beta}{kT}\exp\left(-\frac{U_0}{kT}\right) \tag{5.21}$$

decreasing as the temperature increases. If the temperature of the mantle is high enough for flow to occur as in a liquid, then it may be that the shear stresses are supported dynamically by the movement of the material.

This discussion is of necessity rather vague and general, three factors preventing any greater precision. In the first place, we only know the variations of the acceleration due to gravity at the surface; we do not know the depth at which the corresponding variations of density occur so that we know neither the magnitude of the shear stresses nor the depths at which they are developed. Secondly, there are large uncertainties in estimates of the increase of temperature with depth within the Earth and, thirdly, it is difficult to relate laboratory experiments on the creep of mantle material to the behaviour of the mantle because the time scales for experiment and terrestrial behaviour are so disparate. Lastly, the temperature is probably not independent of creep behaviour of the material for, if the material can move, it carries heat with it and so changes the distribution of temperature in the Earth.

5.6 Dissipation

So far, it has been implicitly assumed that the material of the Earth behaves in a perfectly elastic way, stress being always proportional to strain. Real materials do not behave like that and, although the departures from perfect elasticity within the Earth are in general small, yet they can be detected and some of the effects are very important.

If stress is proportional to strain, then the free vibrations of an elastic system follow the equation of motion

$$\ddot{\theta} + \omega^2 \theta = 0 \tag{5.22}$$

the solution is

$$\theta = \mathrm{Re}\,\theta_0\,e^{i\omega t}$$

In general, more than one coordinate is required to describe the behaviour of an oscillating system but, provided the system is perfectly elastic, then each coordinate will obey a simple harmonic differential equation of motion.

The energy E stored in an oscillating system is proportional to θ^2 ($\sum_i \theta_i^2$ for more than one degree of freedom) and the change of the energy of the system in one cycle is

$$\oint \frac{dE}{dt}\, dt = 2 \int_0^{2\pi/\omega} \theta\dot\theta\, dt \tag{5.23}$$

Now, in simple harmonic motion, $\dot\theta$ is in quadrature with θ—if θ is proportional to $\cos\omega t$, then θ is proportional to $\sin\omega t$—and hence

$$2 \int_0^{2\pi/\omega} \theta\dot\theta\, dt = 0$$

that is to say, in a perfectly elastic system, the system does not lose energy.

Suppose, now, that the differential equation contains a term proportional to $\dot\theta$:

$$\ddot\theta + q\dot\theta + \omega^2\theta = 0 \tag{5.24}$$

The solution is then

$$\theta = \operatorname{Re}\theta_0\, e^{i\gamma t}$$

where

$$\gamma = \omega + \tfrac{1}{2}iq \ \text{ provided } q \ll \omega$$

and then $\dot\theta$ is equal to

$$(i\omega - \tfrac{1}{2}q)\, e^{i\omega t}\, e^{-\frac{1}{2}qt}$$

or

$$i\omega\, e^{-\frac{1}{2}qt}\, e^{i(\omega t + \phi)}$$

where

$$\phi = q/2\omega$$

The change of energy per cycle is proportional to

$$\tfrac{1}{2}\sin\phi = q/4\omega$$

It is usual to denote the ratio of the stored energy to the energy lost per cycle by Q, so that in the simple case of the almost harmonic oscillator

$$Q = 4\omega/q \tag{5.25}$$

The Earth also undergoes forced oscillations, driven particularly by the tidal potential. Suppose θ obeys the equation

$$\ddot{\theta} + \omega^2\theta = F$$

where

$$F = F_0 e^{ivt}$$

then

$$\theta = \frac{F_0 e^{ivt}}{\omega^2 - v^2} \tag{5.26}$$

If now

$$\ddot{\theta} + q\theta + \omega^2\varepsilon = F_0 e^{ivt}$$

then

$$\theta = \frac{F_0 e^{ivt}}{\omega^2 - v^2 + iqv} \tag{5.27}$$

and instead of θ being in phase with the force F it suffers a phase shift ϕ, where

$$\tan\phi = qv/(\omega^2 - v^2)$$

The idea of the Q factor may be extended to travelling waves (stationary waves are covered by the foregoing considerations). If instead of the simple wave equation, a displacement u obeys an equation

$$\rho\ddot{u} + q\dot{u} = \sigma\,\partial^2 u/\partial x^2$$

then

$$u = u_0 \exp\{-i(\omega t - kx)\} \tag{5.28}$$

where

$$k = \kappa_1 - i\kappa_2$$

Provided q is small, $\kappa_1 = \omega/c$, where c, equal to $(\rho/\sigma)^{1/2}$, is the speed of the wave motion, while κ_2 is equal to $(q/2)(\sigma\rho)^{1/2}$.

The energy per unit volume of the solid is the average value of $\rho\dot{u}^2$, or $\tfrac{1}{2}\pi\rho\omega^2 u^2\times e^{-2\kappa_2 x}$ and the rate at which energy crosses a surface parallel to the wave front is $\tfrac{1}{2}\rho c\omega^2 u_0^2 e^{-2\kappa_2 x}$ per unit time.

Now consider a tube of the material of unit cross section and of length equal to one wavelength λ. The energy stored in a volume centred on $x = 0$ is

$$\tfrac{1}{2}\rho\omega^2 u_0^2 \lambda$$

or

$$\pi\rho\omega^2 u_0^2/\kappa_1$$

while the difference between the energy entering and leaving the volume in a time $2\pi/\omega$ is

$$\pi\rho c\omega u_0^2(e^{2\kappa_2\lambda} - e^{-2\kappa_2\lambda})$$

that is

$$4\pi\rho c\omega u_0^2\kappa_2\lambda$$

or

$$8\pi^2\rho c\omega u_0^2\kappa_2/\kappa_1$$

namely

$$8\pi^2\rho c^2 u_0^2\kappa_2$$

Thus Q, the energy stored divided by the energy lost per cycle, is $\omega\rho/4\pi q$, a result analogous to that for the damped oscillator.

The value of Q or q for materials that transmit elastic waves may be found from the decay of amplitude with distance from the source complicated, however, for seismic waves in the Earth by changes of amplitude on reflection or refraction at boundaries within the Earth and by scattering of the waves by irregularities of the material. Estimates of Q both for bodily waves and for surface waves have nonetheless been made.

Q can also be found from the free oscillations of the Earth for, because of dissipation, the energy of the oscillations is spread over a range of frequencies around the peak. If the range of frequencies within which the amplitude exceeds half that at the peak is Δv, while the peak frequency is v_0, then it follows from the equations for free oscillations that

$$Q = v_0/\Delta v$$

Figure 5.12 shows in summary form values of Q for different ranges of period within the Earth. It can be seen that Q, instead of being proportional to frequency as would be expected from the simple models just considered, is almost independent of frequency (to within an order of magnitude) over a very wide range, whether the data come from laboratory experiments on rocks, bodily or surface waves, free oscillations, tidal oscillations or the Chandler wobble, a range of period from less than 1 s to about 1 y.

It is possible to write down an empirical relation between stress, strain and time (Jeffreys, 1970, p. 12) that will give values of Q independent of frequency but, to understand the physical significance of the independence of Q from frequency, it is necessary to understand how solids dissipate elastic energy. It was mentioned before that dislocation in a single crystal end (are 'pinned') on other dislocations or on impurities. If a sufficiently large stress is applied in one direction, the dislocation may move as already described, but if an oscillatory stress is applied the line of the dislocation will bow out like an elastic string between its two ends in the direction of the stress and, when the stress reverses, the bow will be in the opposite direction (figure 5.13). A pinned dislocation, like an elastic string, has a

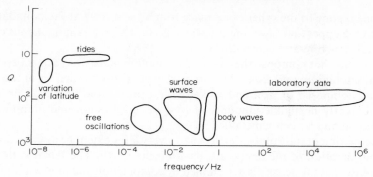

Figure 5.12 Dependence of Q on period

characteristic frequency and so, when an oscillatory stress is applied, the motion of the dislocation will not in general be in phase with the stress and elastic energy will be lost. If all pinned dislocations have comparable characteristic frequencies, the value of Q will depend more or less sharply on frequency; the fact that Q for the Earth is nearly constant over a wide range of frequency implies that the characteristic frequencies of the dislocations (or rather of pinned grain boundaries) also cover a very wide range, a conclusion perhaps not very surprising in view of the range of composition and materials within the Earth.

Figure 5.13 Forced oscillations of a pinned dislocation

5.7 Summary

Let us now try to summarize the picture of the Earth we have established in this chapter, recalling that the density and elastic moduli are derived from the dynamical properties and the elastic behaviour of the Earth, while the interpretation of the data is based on experimental and theoretical studies of silicate minerals at high pressure.

The outer parts of the Earth show great complexity. The crust is composed of a wide range of rocks and minerals and is sharply divided into oceanic and continental types. The mantle below the crust is apparently more uniform but evidence of lateral variation is found in the low harmonic terms in the gravitational potential as well as in seismic studies which show the velocity to be low in such areas as California or below the mid-oceanic ridges.

In the outer parts, the properties of the Earth are determined by the crystal structures of the minerals of which it is constituted, for lattice forces are much greater than the external pressure. At greater depths, lattice forces are small compared with the hydrostatic pressure, and the properties are determined essentially by atomic number, that is, by average chemical composition. The change from one regime to the other takes place roughly speaking at the boundary zone between the upper and lower mantle which is controlled by simultaneous changes of crystal structure and composition.

The atoms that compose the mantle are predominantly oxygen, silicon, magnesium, aluminium and iron; the core is sharply distinguished from the mantle by being composed of iron and nickel with a proportion of silicon or sulphur but apparently negligible amounts of oxygen, magnesium and aluminium, by being liquid and by being metallic.

The equation of state of the Earth can to a very large extent be discussed without any consideration of the temperature within the Earth. The Earth is thus cold, in the sense that the density is determined by composition and pressure and matter is unionized, in contrast with the stars where the temperature has a dominant effect and matter is gaseous and ionized. However, the effects of temperature cannot be entirely neglected; if they could, the Earth would be entirely static and there would be no shearing stresses and no lateral variations for these are incompatible with a structure determined entirely by hydrostatic pressure.

Examples for chapter 5

5.1 Calculate the following moments of inertia about an axis

    (a) of a spherical shell ($\frac{2}{3}Ma^2$)

    (b) of a uniform sphere ($\frac{2}{5}Ma^2$)

(Take spherical polar coordinates with respect to the axis. The moment of inertia of a ring at colatitude θ is equal to the mass of the ring multiplied by the square of the perpendicular distance from the axis.)

5.2 Calculate the flattening f of the Earth according to Darwin's formula (equation (5.18)) from the following data:

$$J_2 = 1{\cdot}0827 \times 10^{-3}$$
$$H = 3{\cdot}275 \times 10^{-3}$$
$$m = 3{\cdot}45 \times 10^{-3}$$

5.3 Calculate the densities at 100 km inside the surface of (a) the top of the mantle, (b) the core, from the following data:

    (a) radius 6330 km

$$\alpha = 7{\cdot}8 \text{ km s}^{-1} \qquad \beta = 4{\cdot}4 \text{ km s}^{-1}$$
$$g = 9{\cdot}8 \text{ m s}^{-2} \qquad \rho = 3350 \text{ kg m}^{-3}$$

    (b) radius 3500 km

$$\alpha = 8{\cdot}1 \text{ km s}^{-1} \qquad \beta = 0$$
$$g = 10{\cdot}40 \text{ m s}^{-2} \qquad \rho = 9500 \text{ kg m}^{-3}$$

Calculate also the masses of shells between 6230 and 6330 km and between 3400 and 3500 km.

5.4 In a shock wave experiment, the initial density is 8000 kg m^{-3}, the speed of the shock wave is 10 km s^{-1} and the speed of the material behind the shock 2·5 km s^{-1}.

Calculate the pressure and density behind the shock front and the change in internal energy per kg, taking the pressure in front of the shock to be zero.

$$\left(\begin{array}{c} p_1 = 2 \cdot 10^{11} \text{ N m}^2, \; \rho_1 = 10\,000 \text{ kg m}^{-3} \\ \Delta E = 2 \cdot 5 \times 10^5 \text{ J kg}^{-1} \end{array} \right)$$

To what temperature would the material have to be heated at atmospheric pressure for the same increase of energy, assuming the heat capacity to be 500 J kg^{-1} $degK^{-1}$.

Suppose that the bulk modulus is linearly related to pressure:

$$K = K_0 + 3 \cdot 5p$$

Ignoring the change of temperature in the shock, determine K_0 from the initial and final pressures and densities.

(Integrate the equation $d\rho/\rho = dp/K$ between the initial and final pressures, substitute the initial and final densities and solve the resulting equation for p/K_0.)

5.5 Calculate the speed of a shock wave, u_s, and the speed of the material behind the shock, given that the initial pressure is atmospheric (10^5 N m^{-2}), the pressure behind the shock is 84×10^{11} N m^{-2}, the initial density is 7700 kg m^{-3} and the final density is 9000 kg m^{-3}.

Calculate also the change in internal energy.

Further reading for chapter 5

Adams, L. H., and Williamson, E. D., 1923, *J. Wash. Acad. Sci.*, **13**, 418–28.

Ahrens, T. J., and Petersen, C. F., 1966, 'Shock wave data and the study of the Earth', in *The Application of Modern Physics to the Earth and Planetary Interiors* (Ed. S. K. Runcorn) (New York, London, Sydney, Toronto: Wiley–Interscience), ix + 692 pp.

Binns, R. A., Davis, R. H., and Reed, S. J. B., 1969, *Nature (Lond.)*, **221**, 943.

Bullen, K. E., 1963, *Seismology* (Cambridge: Cambridge University Press).

Cook, A. H., 1972, 'The dynamical properties and internal constitution of the Earth, the Moon and the planets', *Proc. Roy. Soc., Ser. A*, **328**, 301–36.

Jeffreys, H., 1970, *The Earth*, 5th edn (Cambridge: Cambridge University Press), xii + 525 pp.

Kaula, W. M., 1968, *An Introduction to Planetary Physics* (New York, London, Sydney, Toronto: Wiley), ix + 490 pp.

Kittel, C., 1968, *Introduction to Solid State Physics*, 3rd edn (New York, London, Sydney: Wiley), 648 pp.

Kraut, E. A., and Kennedy, G. C., 1966, 'New melting law at high pressure', *Phys. Rev.*, **151**, 668–75.

Mott, N. F., 1968, 'Metal–insulator transition', *Rev. Mod. Phys.*, **40**, 667–83.

Rice, M. H., McQueen, R. G., and Walsh, J. M., 1958, 'Compression of solids by strong shock waves', *Solid St. Phys.*, **6**, 1–63.

Robertson, E. C. (Ed.), 1972, *The Nature of the Solid Earth* (London: McGraw-Hill).

Wyllie, P. J., 1966, 'High-pressure techniques', in *Methods and Techniques in Geophysics*, Vol. 2 (Ed. S. K. Runcorn) (London, New York, Sydney: Wiley–Interscience), ix + 314 pp.

CHAPTER SIX

Radioactivity and the ages of rocks

'When I do count the clock that tells the time.'
Shakespeare, *Sonnet* XI

6.1 Introduction

Prior to the discovery of radioactivity, geologists had established a relative time scale for rocks at the surface of the Earth and were able to state in what order each was formed. Suggestions had also been made about ways in which ages in years could be estimated, for example, from the content of salt in the oceans and the rate at which it is washed out of the rocks into the seas; or from the total accumulated thickness of sand and mud and the rate at which such materials are deposited; or from the time which the evolution of animal life was thought to require. Lord Kelvin calculated the rate at which the Earth would cool supposing that it had no internal source of heat, and estimated the age on the assumption that the original temperature was that of the Sun. The age he found, 20 My, was very much less than that considered necessary by geologists to accommodate evolution, and led to great controversy. None of the methods is in fact well founded, for they assume that processes now seen to be going on at certain rates have always gone on at those rates, and there are really no grounds for making such an assumption about the distant past when the processes in question depend upon the geology, meteorology and oceanography of the Earth at the time.

The discovery of radioactivity changed the situation completely for it showed there was an internal source of heat in the Earth, invalidating Kelvin's calculations which now give a not very effective lower limit to the age of the Earth, and it provided a means by which the ages of rocks could be found. The essential difference between radioactive decay and other processes which were suggested for estimating the age of the Earth is that radioactive decay depends on nuclear forces involving energies so great that ordinary chemical or atomic forces have no detectable effect upon the rate at which radioactive decay takes place. Whereas the solution of salt, for example, is strongly dependent on such conditions as temperature and rainfall, the rate of radioactive decay may be taken to be independent of terrestrial conditions and so, provided that the rate can be found in the laboratory, ages of rocks can be estimated from the accumulated products of radioactive decay in rocks. Ages were estimated by Boltwood (1907) from the ratio of lead to uranium very shortly after the discovery of radioactivity and, not much later, Lord Rayleigh estimated ages from the ratio of helium to uranium.

The principle on which ages may be estimated from the contents of radioisotopes and their decay products is very simple, but in practice many problems arise. In the first place, the quantities of radioactive isotopes and their products are usually extremely small, save for uranium ores, and the measurement of their abundances requires the most delicate experimental techniques. Secondly, the rates at which the isotopes decay must be found experimentally, again not always a straightforward matter, so that there have been on occasion substantial revisions of the adopted values. The most important factor is, however, the possibility that either the parent or the product material may be lost from or gained by rocks or minerals through diffusion or chemical changes.

6.2 The radioactivity of the Earth

The results of the studies of the physical properties of the Earth set out in the previous chapters show that the Earth is composed of compounds of silicon and oxygen with metals, and that the proportion of oxygen decreases and that of metals increases in going to greater depths. The crust comprises sedimentary rocks with silica, feldspars and clay minerals; and igneous and metamorphic rocks with silica, feldspars, pyroxenes and olivine. The continental crust consists of material of acidic to intermediate composition in the upper part, while in the lower part it probably has a basic composition, similar to that of gabbro. The oceanic crust is denser than the continental crust and, apart from the upper sedimentary layer, is of predominantly gabbroic composition.

Of the elements that compose the major minerals of the crust and mantle, only potassium has a radioactive nuclide, that of mass number 40. The effects of radioactive nuclides in the Earth will therefore be controlled to a considerable extent by the distribution of potassium in rocks; the rocks that contain the most potassium are those that are the more acidic, containing feldspars and micas, and those are the rocks of lowest density that lie near the top of the continental crust but which are scarce or absent deeper in the continental crust and in the crust below the oceans. Sedimentary rocks also contain appreciable quantities of potassium, mainly in minerals derived from the feldspars by hydrolysis.

The other main radioactive nuclides found in rocks are uranium-235 and 238, and thorium-232. The amount of potassium in the more acidic rocks reaches 4%, but uranium and thorium occur in rocks only as trace elements with abundances of a few parts per million or less; however, the abundance of the potassium isotope of mass number 40 is only about 1 part in ten thousand of the total abundance of potassium, so that the quantities of the principal radioactive nuclides are comparable. The distribution of the trace elements is determined by, for example, the fact that uranium has rather volatile compounds, and by the fact that the ionic radii of uranium and radium are relatively large so that they cannot easily be accommodated in a silicate crystal lattice. Thus it is that uranium and thorium, like potassium, are found predominantly in the upper part of the crust of the continents. A further radioactive nuclide, rubidium-87, occurs in trace elements in association with potassium, particularly in micas.

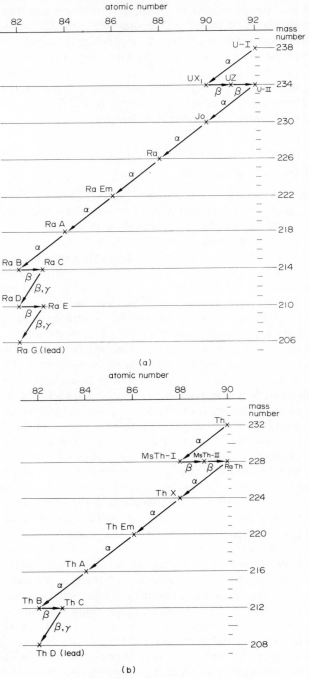

(a)

(b)

Figure 6.1 Decay schemes of U and Th

Table 6.1   Properties of radioactive nuclides in the Earth

Nuclide	Decay scheme	Decay constant (10^{-11} y^{-1})	Half-life (10^9y)	Heat generation (W kg^{-1})
rubidium-87	^{87}Rb $\rightarrow$ ^{87}Sr + β	1·39	49·8	—
potassium-40	^{40}K $\rightarrow$ ^{40}Ar (electron capture)	5·85	11·8	$2·5 \times 10^{-5}$
	$\rightarrow$ ^{40}Ca + β	47·2	1·47	
thorium-232	^{232}Th $\rightarrow$ ^{208}Pb + 6α + 4β	5·0	13·9	$2·5 \times 10^{-5}$
uranium-235	^{235}U $\rightarrow$ ^{207}Pb + 7α + 4β	97·2	0·71	$5·7 \times 10^{-4}$
uranium-238	^{238}U $\rightarrow$ ^{206}Pb + 8α + 6β	15·4	4·5	$9·4 \times 10^{-5}$

The properties of the naturally occurring radioactive nuclides of importance in geophysics are summarized in table 6.1. Rubidium has the simplest scheme of decay, by β decay to strontium-87 and its half-life is the longest at nearly 5×10^{10} y. Potassium-40 undergoes β decay to calcium-40 and also produces argon-40 by electron capture—the latter is the more important process. Uranium and thorium are transformed by sequences of α and β decay to isotopes of lead, as shown in figure 6.1. Table 6.1 also contains values of the heat generated in the radioactive decay of the various nuclides. The internal energies of the parent nuclei are greater than those of the daughter nuclei into which they decay; the difference appears as the kinetic energy of the alpha or beta particles or the photon energy of gamma rays that are emitted. The particles or gamma rays are eventually stopped in the surrounding minerals and heat them up.

Ages estimated from the decay of radioactive nuclei evidently require good values of the decay constants. The general method of measuring small decay rates is to count the rate at which α or β particles or γ rays are emitted from a known mass of the radioactive isotope. If the number of the latter is N, the rate of emission of particles or γ rays will be λN, where λ is the decay constant. When the rates are low, as for long-lived isotopes, it is important to count for a long time to attain good statistical accuracy and care must be taken to avoid counting cosmic rays or other background radiation. On the other hand, if the rays have low energy (for example, the β particles from ^{40}K) then precautions must be taken against missing counts.

6.3   Radioactive dating of rocks and minerals

The fundamental fact about radioactive decay is that the probability of a nucleus decaying at any instant is a constant. If a sample contains N nuclei with decay constant λ (the unit of the decay constant is the reciprocal second), the number that decay each second will be λN. Thus the rate of change of the number of parent nuclei is given by

$$\frac{dN}{dt} = -\lambda N \qquad (6.1)$$

while that of the number of daughter nuclei N' is given by

$$\frac{dN'}{dt} = +\lambda N \qquad (6.2)$$

Suppose that the sample of parent nuclei was assembled at some unique identifiable time, which may then be taken as the origin from which time is measured. The number of parent nuclei at a time t after the instant of assembly will be given by the solution of equation (6.1), namely

$$N = N_0 e^{-\lambda t} \qquad (6.3)$$

where N_0 is the initial number of parent nuclei.

On substituting this value of N in equation (6.2) the differential equation for N' becomes

$$\frac{dN'}{dt} = \lambda N_0 e^{-\lambda t} \qquad (6.4)$$

which has the solution

$$N' = A - N_0 e^{-\lambda t} \qquad (6.5)$$

where A is a constant of integration. Now, initially, the number of daughter nuclei is zero, and so $A = N_0$, or

$$N' = N_0(1 - e^{-\lambda t}) \qquad (6.6)$$

Suppose that, after some time $t_{1/2}$, the number of parent nuclei is just half the original number. Then

$$\exp(-\lambda t_{1/2}) = \tfrac{1}{2}$$

so that

$$\lambda t_{1/2} = \ln 2$$

or

$$t_{1/2} = \lambda^{-1} \ln 2$$

$t_{1/2}$ is known as the *half-life* of the parent nuclide; values for the geophysically important nuclides are given in table 6.1.

If we eliminate N_0 between equations (6.3) and (6.6), we find that

$$N' = N(e^{\lambda t} - 1)$$

whence it follows that

$$e^{\lambda t} = 1 + N'/N$$

or

$$t = \lambda^{-1} \ln(1 + N'/N)$$

a formula that enables the time t since the assembly of the sample to be calculated from the numbers of parent and daughter nuclei.

It has been supposed that we are dealing with the simplest possible decay scheme in which a parent decays to a single stable daughter. That is the case for the decay of rubidium-87 but all the other radioactive nuclides of geophysical importance have more complex decay schemes. Uranium and thorium, as shown in figure 6.1, are the first members of long series of decays that terminate in isotopes of lead. Let N_i be the number of nuclei of the ith member of the chain at time t and let N_{i+1} be the number of the $(i + 1)$th member. The increase in the number of the $(i + 1)$th nuclide in the time t is the difference between the number of the ith nuclide that decay to the $(i + 1)$th and the number of the $(i + 1)$th that decay. The differential equation for the rate of change of N_{i+1} is therefore

$$\frac{dN_{i+1}}{dt} = \lambda_i N_i - \lambda_{i+1} N_{i+1} \tag{6.7}$$

and the numbers of the various nuclides in the chain are found by solving the set of simultaneous differential equations consisting of one equation like (6.7) for each nuclide. The initial condition is that, at the origin from which time is measured, all the N_i are zero except N_1 which has the value N_0.

The general solution to the set of simultaneous equations is complex (Jeffreys, 1970, p. 345) but simplifies very considerably in the conditions that apply to the uranium and thorium chains, in which all the half-lives are relatively short except that of the decay from the initial uranium or thorium, that is to say,

$$\lambda_1 \ll \lambda_2, \cdots \lambda_i, \cdots \lambda_n$$

where n denotes the end member of the sequence.

The general solution then becomes just

$$\left. \begin{aligned} N_1 &= N_0 e^{-\lambda t} \\[2mm] N_i &= \frac{\lambda_1}{\lambda_i} N_1 \\[2mm] N_n &= N_0(1 - e^{-\lambda t}) \end{aligned} \right\} \tag{6.8}$$

The relation between the numbers of the initial and final nuclides is the same as that between parent and daughter for the simple decay, while the numbers of intermediate nuclides bear a constant ratio (equal to λ_1/λ_i) to the numbers of the initial nuclide. Accordingly the time since the formation of the sample of the initial nuclide can be found in just the same way as for the simple decay from the numbers of initial and final nuclides.

The decay of potassium involves a different complication. Potassium-40 can decay by two processes—β decay to calcium-40 and electron capture to argon-40, with respective half-lives of $1\cdot5$ and $11\cdot8 \times 10^7$ y. There is then a slight change in the formula for the age derived from the simple decay scheme.

Let N_1 be the number of nuclei of potassium-40, N_2 that of argon-40, and N_3 that of calcium-40. Let the constants for decay to argon and calcium be respectively λ_2 and λ_3. The rate of change of the number of potassium nuclei is given by

$$\frac{dN_1}{dt} = -(\lambda_2 + \lambda_3) N_1$$

since potassium nuclei are lost by two processes.

The number of potassium nuclei is therefore

$$N_1 = N_0 \exp\{-(\lambda_2 + \lambda_3) t\}$$

where N_0 is the initial number at the formation of the sample.

Argon nuclei are formed only through the decay with constant λ_2 and their number therefore increases at the rate given by

$$\frac{dN_2}{dt} = \lambda_2 N_1 = \lambda_2 N_0 \exp\{-(\lambda_2 + \lambda_3)t\}$$

Then

$$N_2 = A - \frac{\lambda_2}{\lambda_2 + \lambda_3} N_0 \exp\{-(\lambda_2 + \lambda_3)t\}$$

and applying the condition that, when t is zero, so is N_2 (and N_3) it follows that

$$N_2 = \frac{\lambda_2}{\lambda_2 + \lambda_3} N_0 [1 - \exp\{-(\lambda_2 + \lambda_3) t\}] \tag{6.9}$$

Hence the ratio of argon to potassium nuclei is

$$\frac{N_2}{N_1} = \frac{\lambda_2}{\lambda_2 + \lambda_3} [\exp\{(\lambda_2 + \lambda_3) t\} - 1]$$

and the time since the formation of the sample of potassium is

$$\frac{1}{\lambda_2 + \lambda_3} \ln\left(1 + \frac{\lambda_2 + \lambda_3}{\lambda_2} \frac{N_2}{N_1}\right) \tag{6.10}$$

Both decay constants are required in order to determine times from the ratio of argon to potassium.

The practical application of the elementary ideas of radioactive decay to the determination of the ages of rocks requires careful consideration of four factors. In the first place, the numerical effectiveness of the method depends on being able to find nuclides with decay constants of the same order of magnitude as the times which it is desired to estimate. In fact, the presence of radioactive isotopes in appreciable quantities ensures that this condition is satisfied, for short-lived isotopes no longer exist in the rocks of the Earth and those that are still found there must have lives comparable with the ages of the rocks in which they are found.

The next, and probably the most important, factor is that the idea of an original pure sample of the parent nuclide must be meaningful. If such a sample never existed, then the instant from which times are measured is indefinite. Uranium is often found concentrated in minerals which have been carried upwards in fluids during the final stages of solidification of a rock, but along with it there may be carried lead of radioactive origin, so that the quantity of lead found in the minerals at a later date will be greater than if it were all formed from uranium starting at the time of solidification of the rock. Argon, on the other hand, being a gas, is not retained in minerals until they are quite cool, so that the idea of an initial pure sample of potassium is likely to be a better one than that of a pure sample of uranium. However, argon may escape from a mineral after initial crystallization if it should later be heated, so that the age estimated from an argon–potassium ratio is that of the last time at which a rock was hot enough to allow argon to escape. Thus, the circumstances in which rocks were formed must be carefully examined before the significance of the ages determined radio-actively can be properly understood.

The third requirement for a meaningful estimate of age is that the daughter nuclei should have accumulated only through the radioactive decay—there should have been none present initially, none should have been added by other means and none taken away. Daughter atoms may be lost by gaseous diffusion, as for argon, especially if the rocks or minerals are disturbed in any way, either naturally or artificially; and they may be lost by solution from the rocks. Atoms can diffuse between different types of mineral as well as be lost to the rock entirely, and thus it is that ages determined from separated minerals may differ from each other and from an age determined from the average abundances of nuclides in the rock as a whole.

It is also possible for daughter atoms to be brought into a rock or mineral from outside sources, so making the apparent age too great, while in certain cases the daughter nuclides also occur originally, that is they were formed through nuclear processes in the early history of the Earth independently of the decay from which they now arise. Both strontium and lead daughter nuclides occur in this way and the problem will be considered further below in describing the rubidium–strontium method.

The fourth requirement for a satisfactory determination of age is that it should be possible to measure the abundance of the various nuclides with adequate accuracy. Prior to the development of mass spectrometers, chemical methods had to be employed and the precision was poor, but nowadays most measurements are made with mass spectrometers and achieve high sensitivity and precision.

The potassium–argon system is from all points of view the simplest. The event of which the age is found is usually unambiguously that at which the rock was last hot enough to lose any argon previously in it. The abundance of potassium-40 is a constant fraction of the stable isotopes which are abundant enough in the more acidic rocks for spectroscopic methods to be used to determine them. The intensity of light from potassium emitted by a flame fed with a solution prepared from the rock is compared with that from a solution of known concentration and so the quantity of potassium in the rock is found by comparison. The uncertainty is about 2 %. The flame photometer can be used with the relatively large amounts of potassium in acid rocks; for rocks with lower concentrations, the potassium is separated chemically by ion exchange or otherwise and then measured with a mass spectrometer, using for comparison a sample with known concentrations of ^{41}K and ^{40}K. The argon is measured by expelling it all from the rock at high temperature and low pressure, adding to it a known quantity of non-radiogenic argon, and then measuring the ratio of radiogenic to non-radiogenic argon in the mixture. Precautions must of course be taken to ensure that argon is not lost in the course of the manipulations. The minerals may take up argon by diffusion from the atmosphere and there will also be some atmospheric argon in the apparatus and it is essential to determine how much is present and to correct for it. The bulk of atmospheric argon is argon-40 (derived from radioactive sources) but there are in addition small quantities of argon-36 and 38 which are not derived from potassium; by measuring the quantity of one of these isotopes in the material, the quantity of argon that has diffused in from the atmosphere may be estimated. Minerals differ in the ease with which they lose argon. Some feldspars lose argon even at room temperature and so ages determined from them are not reliable; the feldspar, sanidine, however, which is formed at high temperature, retains argon well. Micas lose argon at about 200 °C and ages found from them thus indicate the time at which the rock last cooled to ambient temperature. So small are the quantities of argon that can be determined with the mass spectrometer, that, although the half-life is $11 \cdot 8 \times 10^9$ y, ages as low as $1 \cdot 4$ My have been determined (McDougall, 1966).

Rubidium is a trace element in nature, with abundances much less than those of potassium, and its half-life is nearly five times as great as that of potassium, so that very small quantities of strontium have to be determined. Having the longer half-life it is more suitable for the greater ages. Strontium is not lost as argon is, by diffusion at low temperatures, nor is it absorbed from the atmosphere. However, the daughter strontium isotope occurs in original form independently of the presence of rubidium in a rock, and allowance for the original strontium must be made in estimating ages. The average ratio of rubidium to strontium in the crust of the Earth is quite large, about 1:4, and, if that were to be maintained in all minerals, it would be very difficult to determine the radioactive strontium in the presence of that occurring naturally. Fortunately, because of the different chemistry of strontium and rubidium, rubidium occurs preferentially in many minerals such as micas and potassium feldspars, and the radiogenic strontium can be determined satisfactorily in the presence of original strontium.

In addition to the isotope of mass 87, strontium isotopes of masses 84, 86 and 88 occur naturally independently of rubidium. Suppose that the original ratio of 87 to 86 is known, 86 being a convenient standard of comparison; then the quantity of mass 87 in the mineral arising from radioactive decay is equal to Qr, where Q is the quantity of mass 86 and r is the ratio of radiogenic 87 to 86 given by

$$r = \left(\frac{\langle^{87}\text{Sr}\rangle}{\langle^{86}\text{Sr}\rangle}\right)_{\text{measured}} - \left(\frac{\langle^{87}\text{Sr}\rangle}{\langle^{86}\text{Sr}\rangle}\right)_{\text{original}}$$

(The brackets $\langle\,\rangle$ denote the atomic abundances of the respective nuclides.)

If the rubidium is much more abundant than the strontium, it is sufficient to take for the original ratio the value for rocks rich in strontium, namely, about 0·71; where that assumption is not adequate, it is possible to determine the original ratio by comparing abundances in different minerals in the rock. It is supposed that the material from which the rock was formed had a certain initial ratio of strontium to rubidium and that, in the course of formation, rubidium and strontium were exchanged between minerals, giving rise to different original ratios. Each sample will then give a separate equation

$$\left(\frac{\langle^{87}\text{Sr}\rangle}{\langle^{86}\text{Sr}\rangle}\right)_{\text{measured}} = \left(\frac{\langle^{87}\text{Sr}\rangle}{\langle^{86}\text{Sr}\rangle}\right)_{\text{original}} + \left(\frac{\langle^{87}\text{Sr}\rangle}{\langle^{86}\text{Sr}\rangle}\right)_{\text{radiogenic}}$$

but they are all related because

$$\langle^{87}\text{Sr}\rangle_{\text{radiogenic}} = \langle^{87}\text{Rb}\rangle\,(e^{\lambda t} - 1)$$

where λ is the decay constant of ^{87}Rb. Hence

$$\left(\frac{\langle^{87}\text{Sr}\rangle}{\langle^{86}\text{Sr}\rangle}\right)_{\text{measured}} = \left(\frac{\langle^{87}\text{Sr}\rangle}{\langle^{86}\text{Sr}\rangle}\right)_{\text{original}} + \left(\frac{\langle^{87}\text{Rb}\rangle}{\langle^{86}\text{Sr}\rangle}\right)(e^{\lambda t} - 1)$$

If the ratios $\langle^{87}\text{Sr}\rangle/\langle^{86}\text{Sr}\rangle$ and $\langle^{87}\text{Rb}\rangle/\langle^{86}\text{Sr}\rangle$ for different minerals and for the rock as a whole are plotted against each other as in figure 6.2 they should lie close to a straight line, the *isochron*, the slope of which gives the age of the rock as a whole, whilst the intercept on the ordinate gives the original value of

$$\langle^{87}\text{Sr}\rangle/\langle^{86}\text{Sr}\rangle$$

An instructive example of the effect of heating upon the ages estimated by different methods is given by Tilton and Hart (1963). A crystalline rock about 1300 My old in Colorado (U.S.A.) was intruded by granite some 50 My ago; as a consequence the older rock was heated to a temperature that decreased away from the contact as shown in figure 6.3(a). Ages were determined by the potassium–argon and rubidium–strontium methods upon different minerals as indicated in figure 6.3(b). Argon and strontium have both been removed from the coarse biotite by heating, whereas argon has been retained in the hornblende even quite close to the hot granite. Within the biotite, potassium–argon gives the lower ages because the argon is lost more readily than strontium. Ages determined on feldspars often behave in the erratic way seen in the figure.

The first estimates of the ages of rocks from radioactivity used the uranium and thorium sequences. The α particles accumulate as helium, but helium diffuses readily from minerals so that ages based on helium content are by no means so reliable as those based on argon content. Lead isotopes, like those of strontium,

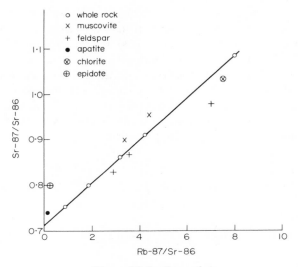

Figure 6.2 Isochron plot

occur originally in a mineral or rock as well as being derived from radioactive decay of uranium and thorium, and similarly the quantity of original lead may be estimated from the quantity of an isotope that is not produced from uranium or thorium, namely lead-204. Because uranium, thorium and lead all have compounds with high vapour pressures, they are often found to be concentrated in minerals which have crystallized during the latest stages of the cooling of an igneous rock, in the minerals in which are often to be found the richest concentrations not only of lead and uranium but of other rare metals as well. The meaning of ages determined on minerals crystallizing late in the solidification of a set of rocks should be considered carefully in each particular case.

The two uranium–lead decay sequences both start with uranium and end with lead so that, if any chemical process has altered the abundances of the parents or the daughters, it should have affected both chains equally. The decay constants are, however, not the same, and it therefore turns out to be possible to detect whether lead has been lost from the system. If no lead is lost, a plot of the ratio $\langle^{206}Pb\rangle/\langle^{238}U\rangle$ against the ratio $\langle^{207}Pb\rangle/\langle^{235}U\rangle$ would give a curve which is convex upwards because the half-life of the uranium-238 system is much greater

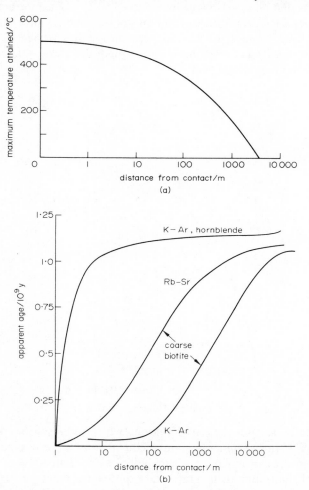

Figure 6.3(a), (b) Ages in the neighbourhood of an intrusion in Colorado (Tilton and Taylor, 1963)

than that of the uranium-235 system, so that the rate of production of lead-207 is much greater than that of lead-206 early in the history of a rock or mineral. The resulting curve is known as a *Concordia plot* and is shown in figure 6.4. Suppose that a mineral were originally formed at time t_1. Then its present isotope ratios would define a point on the Concordia curve. Suppose also, that at a later time t_2, it was altered in some way so that it lost all its lead but none of its uranium. Its present isotope ratios would then define a later point labelled on the curve and, if not all the lead were lost, the characteristic point would lie somewhere on the straight line joining the points at t_1 and t_2 on the concordia curve. It sometimes happens that a group of rocks or minerals in the same region is found to give isotope ratios that lie at various points on the line $t_1 t_2$. The interpretation is that different systems have lost different amounts of lead in the course of the event that occurred at time t_2 and, from the intercepts of the line defined by the observed ratio with the concordia curve, the age of origin of the rocks in the region and the age of alteration may be read off.

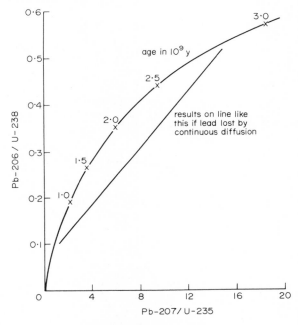

Figure 6.4 Concordia plot

α and β decay are not the only modes by which radioisotopes decay; a small fraction of ^{238}U nuclei undergo spontaneous fission into a number of daughter nuclei having masses much less than half the uranium mass. The fragments into which the uranium disintegrates have quite large kinetic energies and cause considerable damage to the crystals in which they are formed. If a polished surface of such a crystal is etched with acid, the short tracks of the fragments that intersect the surface are picked out and can be counted under a microscope.

Let N_f be the number of tracks in a given volume and $\langle ^{238}U \rangle$ the number of uranium-238 nuclei. Let λ_f be the decay constant for fission and λ_α that for α decay of uranium-238. Then, since λ_α is very much greater than λ_f (table 6.1),

(see p. 150)

$$\frac{\lambda_f}{\lambda_\alpha} \frac{N_f}{\langle ^{238}U \rangle} = e^{\lambda_\alpha t} - 1$$

Because the number of tracks counted is that crossing a surface in the crystal, it is not a simple matter to find N_f, but the count can be calibrated by producing additional tracks in the crystal. Uranium-235 undergoes fission when it is irradiated with neutrons and, if the mineral is irradiated with a known flux of neutrons from a nuclear reactor, then the number of nuclei undergoing fission can be calculated from the flux of neutrons, F, the cross section for fission, σ, and the number of uranium-235 nuclei present. If n_f' is the number of fission tracks counted per unit area from the uranium-235

$$n_f' = K F \sigma \langle^{235}U\rangle$$

where $\langle^{235}U\rangle$ is the concentration of uranium-235 per unit volume and K is a geometrical factor relating the number of tracks per unit volume to the number counted at a surface. Now the ratio of the abundances of uranium-238 and 235 is found to be a constant in all natural materials and so we may put

$$\langle^{238}U\rangle = k\langle^{235}U\rangle$$

where k is a known constant.

Hence

$$\langle^{238}U\rangle = kn_f'/KF\sigma$$

But, if the number of natural tracks per unit area is n_f, the number N_f of natural tracks per unit volume is equal to n_f/K.

The time since fission tracks started to accumulate in the mineral is then given by

$$e^{\lambda_\alpha t} = 1 + \frac{\lambda_f}{\lambda_\alpha} \frac{F\sigma}{k} \frac{n_f}{n_f'}$$

The fission track method is remarkably sensitive, ages as short as 0·3 My having been determined on glassy rocks containing only 1 part per million of uranium, but fission tracks are annealed at quite low temperatures and so the ages that are estimated are those since the rock was last heated above ambient temperature.

The reliability of ages estimated from concentrations of radioactive isotopes and their products depend critically upon the ease with which isotopes can move into or out of the rocks or minerals, and it is possible to classify the different methods and the material upon which they are employed according to the consistency between methods and the overall reliability of the results. Such a classification is given in table 6.2.

The foregoing methods are applicable to the oldest rocks found at the surface of the Earth, but in general are not effective for the youngest rocks because the quantities of daughter products are then very small, even though by very delicate methods of analysis it has been possible to estimate ages of less than 1 My. Nonetheless, isotopes with half-lives comparable with the ages of the oldest rocks found at the surface of the Earth are essentially unsuitable for dating the youngest

Table 6.2   Systems of radioactive dating

Potassium–argon	Gives most recent metamorphism (mica)
	Applicable to ages from 1 to 4000 My
Rubidium–strontium	Gives age of formation of rock (whole rock analyses)
	Minerals may interchange isotopes
	Suitable for acidic igneous rocks
Uranium–lead	Lead can be lost by diffusion
	Can be used to determine age of crust
Uranium–fission	For glassy materials
	Very sensitive

rocks, for which it would be better to use isotopes with short half-lives. Such isotopes must be being produced at the present time or they would not now be present in significant amounts. A radioactive isotope of carbon meets this condition. When nitrogen-14 nuclei in the atmosphere are irradiated by neutrons from cosmic rays they are transformed into radioactive carbon-14 nuclei:

$$^{14}N + n \rightarrow {}^{14}C + p$$

The ^{14}C nuclei are oxidized to carbon dioxide, mix with the atmosphere and are taken up by plants or animals, where they then undergo a β decay to ^{14}N. The age of material such as wood or the shell of an animal made of calcium carbonate can therefore be found from the quantity of radioactive carbon that it now contains. The carbon is determined by converting the material to carbon dioxide and counting the number of β rays emitted by the gas, a delicate operation because the β rays have very low energy and because the quantities of radioactive carbon are very small. Ages up to about 50 000 y can be found in this way; because the parent rather than the daughter isotope is measured, the method is more accurate for the shorter ages.

6.4   The geological time scale

The ages of only relatively very few specimens of rocks or minerals have been estimated by radioactive methods and the ages of the great body of rocks at the surface of the Earth are obtained from a time scale that has been set up in a relative way by geological methods and calibrated in terms of years by the few rocks that have been dated radioactively.

The geological time scale is based on the principle that, of two distinct undisturbed sedimentary rocks lying one above the other, the upper one is the younger. Layers of sedimentary rocks are not continuous over the Earth, and so some way must be found to relate a sequence of rocks in one area to that in another. Fossils provide the means of so doing. Fossils are the buried remains of the hard parts of animals, for example, the shells of oysters or the skeletons of fishes, and it was found quite early in the history of geology that one type of fossil was always found later than another whenever the two occurred in the same locality. Some fossils, like living plants or animals, occur only in restricted regions, but others seem to have lived all over the world, and it was realized that the occurrence of a particular fossil in a sedimentary rock meant that that rock must have been formed within a specific span of time. Some types of plants or animals seem to have lived for relatively short times, while others existed for very long periods but, by considering not just individual fossils in a rock but the whole assembly of organisms to be found there, it is possible to subdivide the relative time scale within which the sedimentary rocks were formed to quite fine units, sometimes to less than 1 My.

Igneous rocks contain no fossils but it is usually possible to date them within the fossil time scale. A lava will often be found to have flowed over a rock containing fossils (and therefore of a known age) and to be itself covered with other rocks containing fossils. The age of the lava must then be between the ages of the two fossiliferous rocks above and below it. Other igneous rocks, granites and gneisses, for example, have crystallized some distance below the surface but again, by considering the rocks into which they are intruded, it is often possible to assign them a place in the fossil scale with some precision.

If now the ages of a few rocks accurately placed in the fossil scale are measured absolutely in years by radioactive methods, then it is possible to assign dates to the occurrence of key fossils and so to the successive steps in the fossil scale. The result of such assignments are shown in table 6.3, which contains the names of the major geological subdivisions*, some of the characteristic fossils, and the ages determined radioactively.

It is important to realize how short, relatively speaking, is the period of time from which well-preserved fossils are recovered, that is to say, from the Cambrian onwards. More primitive fossils are found in older rocks, and the complex

Table 6.3   The geological time scale

Era	Period	Start of period (My)	Duration (My)
Cainozoic	Quaternary	1	1
	Tertiary	65	64
Mesozoic	Cretaceous	135	70
	Jurassic	190	55
	Triassic	225	35
Palaeozoic	Permian	270	45
	Carboniferous	340	70
	Devonian	400	60
	Silurian	430	30
	Ordovician	500	70
	Cambrian	600	100
Precambrian		back to 4500	

* The main subdivisions are often named after regions in which they were first studied, for example, the Cambrian in North Wales, the Silurian from the region of the ancient Silures tribe in Wales and the Jurassic from the Jura mountains of France and Switzerland.

development of the Cambrian fossils themselves suggests that they were preceded by a long period in which living organisms existed, although of a type not easily preserved in rocks. The older *pre-Cambrian* rocks, as they are known, have also in general undergone more or less drastic alteration and recrystallization as the result of being buried under later rocks and of having been subject to pressure and temperature in the formation of mountains; they consist for the most part of metamorphic rocks that would not be expected to contain recognizable fossils. It is nonetheless possible to trace types of rock over stretches of countryside and so to set up a relative time scale within a given region, although in the absence of fossils the results may be unreliable. The ages of pre-Cambrian rocks, being

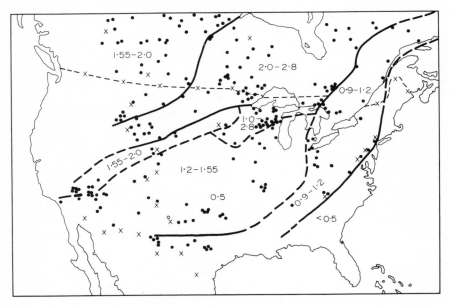

Figure 6.5 Pre-Cambrian ages in North America in 10^9 y (after Tilton and Hart, 1963)

great, can be found with good precision by radioactive methods, and so it has been possible to establish a time scale in the pre-Cambrian although with coarser subdivisions that may not be so consistent between regions as are those in the Cambrian.

Studies of the pre-Cambrian of the North American continent have given an interesting result, for it has been found that the oldest rocks are found around Hudson's Bay, and that successively younger rocks are disposed around this nucleus at increasing distances, as shown in the map of figure 6.5. The oldest rocks have ages of about 3000 My, and the central parts of other continents also appear to be the oldest parts and to have similar ages. The result suggests that the continents were originally much smaller than they are now and have grown outwards by the formation of continental from oceanic crust.

6.5 The age of the crust

The greatest ages found from direct determinations on rocks are about 3500 My, but a study of the isotopes of lead indicates that there was some major mixing of the materials of the crust of the Earth that can be considered to be the time at which the crust as we now know it became stabilized. The time of the event can be established by considering the abundances of the radiogenic isotopes of lead in relation to the non-radiogenic isotope, lead-204.

Let the abundances of all isotopes of lead or uranium be expressed as multiples of the abundance of lead-204 and write

$$x = \langle {}^{206}\text{Pb}\rangle / \langle {}^{204}\text{Pb}\rangle$$

$$y = \langle {}^{207}\text{Pb}\rangle / \langle {}^{204}\text{Pb}\rangle$$

$$u = \langle {}^{238}\text{U}\rangle / \langle {}^{204}\text{Pb}\rangle$$

$$v = \langle {}^{235}\text{U}\rangle / \langle {}^{204}\text{Pb}\rangle$$

Consider a sample of lead in a rock that was formed t_1 years ago (as determined by the techniques described in section 6.3). Suppose that the original lead was formed from a crustal stock that originated t_0 years ago and that at that time the abundances of ^{206}Pb and ^{238}U were x_0 and u_0. Let the present abundances be x_1 and u_1; then x_1 is the sum of the amount of lead originally present plus that produced radioactively before the crystallization of the rock, plus that produced radioactively since crystallization:

$$x_1 = x_0 + u_0\{1 - e^{-(t_0 - t_1)\lambda_{238}}\} + u_1(e^{\lambda_{238} t_1} - 1)$$

There is a similar equation for ^{207}Pb and ^{235}U and such pairs of equations may be written down for rocks of different ages in the same region for which it may be supposed that the age of the crust has significance.

There is a further equation for each rock which follows from the fact that the present ratio $\langle {}^{238}\text{U}\rangle / \langle {}^{235}\text{U}\rangle$ is found very generally to be 139 to 1. Thus

$$\frac{u_0}{v_0} = 139\,\frac{\exp(t_0 \lambda_{238})}{\exp(t_0 \lambda_{235})}$$

Sets of equations such as these may be solved by the method of least squares and the age of the crust of the Earth is estimated to be about 4500 My.

By including lead isotopes measured in meteorites, it is possible to estimate a time t_0 when material of the Earth and meteorites had a common origin. This time is about 4550 My and, as near as any, can be taken as the age of the Earth.

6.6 Ages on the Moon

The samples of the lunar surface brought back by the astronauts of the *Apollo* expeditions contained radioactive isotopes from which the ages could be found. Rubidium–strontium and potassium–argon ages have been obtained for a number of specimens and cluster around 3700 My. In addition, the data on lead isotopes can be combined with those from the Earth and meteorites to give an age for the common parent material of the three bodies. Figure 6.6 shows the lead isotope ratios for two lunar samples plotted on a common diagram with results from meteoritic and terrestrial samples, from which it is concluded that the common age is about 4700 My.

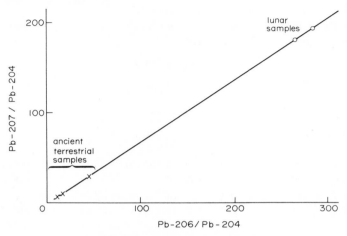

Figure 6.6 Lunar lead isotope ages

Examples for chapter 6

6.1 Calculate the decay constant of potassium −40 by β-decay from the following data:

No. of β particles emitted from ordinary potassium per second: 28 g^{-1}
abundance of ^{40}K in ordinary potassium: $1 \cdot 2 \times 10^{-4}$
Avogadro's number: 6×10^{26} atoms in 12 kg of ^{12}C

Compare your value with that in table 6.1.

6.2 Calculate the age of a sample of mica given that the ratio of the number of nuclei of ^{40}A to the number of ^{40}K is 0·08.

Further reading for chapter 6

Apollo 11, 1970, *Lunar Science Conf.*, reprint from *Science*, **167**, 447–804.

Boltwood, B. B., 1907, *Am. J. Sci.*, **23**, No. 4, 77–88.

Jeffreys, H., 1970, *The Earth*, 5th edn (Cambridge: Cambridge University Press), xii + 525 pp.

McDougall, I., 1966, 'Precision methods of potassium–argon isotopic age determination on young rocks', in *Methods and techniques in geophysics*, Vol. 2 (Ed. S. K. Runcorn) (London, New York, Sydney: Wiley–Interscience), ix + 314 pp.

Tilton, G. R., and Hart, S. R., 1963, 'Geochronology', *Science*, **140**, 357–66.

CHAPTER SEVEN

The temperature within the Earth

'Lasciate ogni speranza, voi ch'entrate.'
Dante, *Inferno*, III

7.1 Introduction

The temperature of the surface of the Earth is fixed by the balance between the heat gained by the ground and the lower atmosphere from the Sun's radiation and that lost by radiation to space. On going down a mine or tunnel towards the centre of the Earth, the temperature is found to rise, showing that heat is flowing from the interior of the Earth outwards. The rate of the flow can be estimated from the rate of increase of temperature together with the thermal conductivity of the rocks near the surface; the first measurements were confined to mines and bore-holes on land but in the last quarter of a century measurements have been made at sea as well, with the somewhat surprising discovery that heat flows out of the Earth below the oceans at very much the same rate as it does from below the continents. The result is surprising because it has seemed that much of the heat comes from radioactive material in the Earth, and the crust below the continents seems to be more radioactive than that below the oceans. The situation is, however, more complex than such elementary ideas suggest: we have in the first place no very firm ideas about the sources of heat within the Earth while, in the second place, the mode of transport of heat within the Earth is not just that of simple conduction. There is then considerable uncertainty about the temperatures within the Earth, but it is not possible to ignore the problems that they present for, undoubtedly, the temperature must have a most important influence upon the behaviour of the Earth and upon the state of material within it.

7.2 The flow of heat to the surface of the Earth

By the definition of thermal conductivity, κ, the rate at which heat flows across unit area within a rigid body is given by

$$Q = \kappa \, dT/dz$$

where Q is the rate of heat flow, T is the temperature and z is a coordinate measured at right angles to the surface across which the heat flow is to be calculated. The units of Q are watts per square metre (W m^{-2}), those of T are degrees Celsius or absolute, and the units of κ are therefore watts per metre per degree (W m^{-1} degC^{-1} or W m^{-1} degK^{-1}).

The rate of flow of heat out of the Earth is about 0·06 W m^{-2} and the conductivity of rocks near the surface is about 2 W m^{-1} degK^{-1}, so that gradients of

some 0·03 degK m⁻¹ are found below the surface of the continents, corresponding
to a rise of 10 degK on going down 300 m. Changes of temperature of that order
may be measured with ample accuracy with mercury-in-glass thermometers and
all the early measurements of gradients were made in that way either in mines or
in boreholes. To observe in a borehole, a thermometer that records the maximum
temperature attained is lowered into the hole and then brought back to the
surface to be read. Because boreholes are filled with water or mud, the pressure
at the bottom is much greater than that at the surface and thermometers are

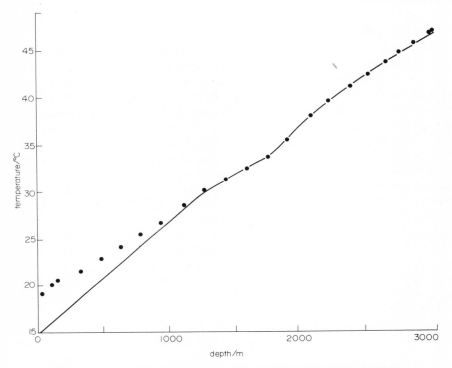

Figure 7.1 Temperature gradient in a borehole in Africa (Bullard, 1939)

encased to make them insensitive to changes of pressure. Nowadays a set of
electrical thermometers, often thermistors, is attached to a cable and lowered
into the borehole. The rocks below the surface of the Earth are mostly far from
uniform and the thermal conductivity may vary quite rapidly. It is therefore
necessary to know the types of rock across which the temperature is measured,
to know the conductivities of all the main types of rock that the hole passes
through and to measure the temperature at least at every boundary between
rocks of significantly different conductivity. Temperature gradients can vary
considerably within a borehole (figure 7.1) and, if only the top and bottom
temperatures were to be measured, quite misleading results might be obtained.

The heat flowing out from the interior of the Earth may not be the only source of heat in a borehole and, in particular, the rocks are heated by the work of boring the hole. Care must therefore be taken that all such heat has been dissipated before the temperatures are measured. On the other hand, heat may be removed from the hole. Mines, for instance, are not very suitable for heat flow measurements because the ventilating system cools the rocks, and similarly deep tunnels, whether for traffic or for water, are useless, although heat flow has been successfully estimated from records of the temperatures encountered during the construction of Alpine tunnels. Water circulating naturally in the rocks through which a borehole passes may also heat or cool them and should be looked for carefully if the hole passes through porous rocks such as sandstone or limestone.

The temperature at the surface of the Earth is not constant, varying as it does, in daily, annual and even longer cycles. The effects of these cycles extend downwards into the ground and must be allowed for in calculating the heat flow from the observed gradients. The situation is shown in figure 7.2. The temperature at the surface of the Earth is supposed to oscillate about some mean temperature,

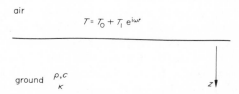

Figure 7.2 The effect of thermal cycling on measured temperatures

T_0, with an amplitude T_1. Let the depth be z measured downwards from the surface. When z is very great, the effects of the variation of temperature at the surface will have died out and the gradient will be Q/κ. Within the ground the temperature satisfies the one-dimensional equation of conductivity:

$$\frac{\partial^2 T}{\partial z^2} = \frac{\rho C}{\kappa} \frac{\partial T}{\partial t} \tag{7.1}$$

where ρ is the density and C the heat capacity. Let the speed of the oscillations at the surface be ω, so that the (complex) surface temperature is

$$T_0 + T_1 e^{i\omega t}$$

The equation of conduction will be satisfied by

$$T = \frac{Qz}{\kappa} + T_1 e^{-\gamma z} e^{i\omega t} \tag{7.2}$$

with

$$\gamma = (1 + i)\left(\frac{\omega \rho C}{2\kappa}\right)^{1/2}$$

so that

$$T = \frac{Qz}{\kappa} + T_1 \exp\left\{-\left(\frac{\omega\rho C}{2\kappa}\right)^{1/2} z\right\} \exp\left[i\left\{\omega t - \left(\frac{\omega\rho C}{2\kappa}\right)^{1/2} z\right\}\right] \qquad (7.3)$$

This means that an oscillation at the surface decays to $1/e$ of the surface amplitude in a depth λ equal to $(2\kappa/\omega\rho C)^{1/2}$.

To estimate some values of λ, ρ may be taken to be 2500 kg m^{-3}, κ to be 2 W m^{-1} degK^{-1} and C to have the typical value for rocks of 10^3 J kg^{-1} degK^{-1}. Then $2\kappa/\rho C$ is about $1{\cdot}6 \times 10^{-6}$ m^2 s^{-1}.

For daily variations of temperature, ω is 2π divided by 1 day, or 7×10^{-5} s^{-1}, and so λ is about 14 cm. For annual variations, λ is greater by the factor $(365)^{1/2}$, that is, it is about 3 m.

The effects of the variations in atmospheric temperature will thus be negligible below about 10 m and will be eliminated if observations of temperature start at greater depths.

The foregoing argument supposes that T_1 is not greater than the range of temperature encountered in the borehole. That is so for holes at the surface of the Earth, but an experiment to measure heat flow on the Moon encounters the difficulty that the monthly cycle of surface temperature is very much greater than the increase of temperature in a hole a few metres deep.

The effects of the longer cycles of temperature may be more disturbing and cannot be estimated so readily. The formation of ice caps in temperate latitudes and their subsequent melting has led to variations of the surface temperature from somewhat less than 0 °C to more than 10 °C in times of the order of 100 000 y; the effects of such variations would not decay to e^{-3} of the surface amplitude in depths of less than 3 km, so that it is likely that all measurements of heat flow in countries that, like Britain, have been subject to an ice age and now enjoy a temperate climate are affected by the cooling of the surface under the ice cap. Areas which have never suffered an ice age or those where the surface temperature does not at present rise much above 0 °C, should give more reliable results.

A further complication in the measurement of heat flow on land is that the gradient will be disturbed if the ground is not level. Thus, if the top of a borehole lies at the summit of a hill, it will penetrate isotherms that are depressed towards those below the surrounding lower ground, while conversely, if a borehole is sited in a valley, it will penetrate isotherms that are raised towards those for the higher ground in either side (figure 7.3).

It is usually a straightforward matter to measure the conductivity of rocks by the divided bar method, the conductivity of a specimen of rock being compared with that of a reference material such as fused silica (figure 7.4). The main problem is to ensure that sufficient specimens are measured to give a truly representative sample of the rocks through which a borehole passes and samples must if at all possible be available from the actual hole in the form of cores taken when the hole was drilled.

The measurement of heat flow at sea is in some ways more difficult because the operations must be conducted from a ship, but it is also in some ways easier and, in fact, there are now far more measurements of heat flow at sea than on land. Only quite recently has it been possible to bore deep holes from ships in the deep oceans, but deep boreholes are not so necessary for heat flow measurement at sea as on land because the daily and annual variations of temperature at the sea bed are far less and so the temperature is less disturbed at moderate depths. Gradients are in fact measured at sea by driving probes into the bottom mud to depths of some 5 m. The measurement is made easier by the fact that the conductivity of mud at the bottom of the sea is not much greater than that of water and is

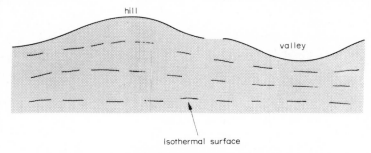

Figure 7.3 Isotherms under a hill and a valley

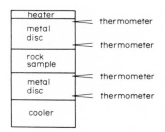

Figure 7.4 The divided bar apparatus for conductivity measurements

less than that of typical hard rocks, so that the gradients of temperature ($0 \cdot 05$ degK m^{-1}) are greater than those measured in boreholes on land; even so, the differences of temperature to be measured are only about $0 \cdot 2$ degK.

Two types of probe have been devised. In the pioneer work of Sir Edward Bullard, a plain cylindrical probe with thermocouples or thermistors at the top and bottom was driven into the bottom mud by a weight fixed to the top of the probe (figure 7.5). The work done when the probe is forced into the mud heats it up by about $0 \cdot 1$ degK and it is necessary to wait for 30 to 40 min before taking the final readings of temperature difference. Even so, the probe will not have attained the temperature of the mud, and a correction, calculated theoretically, has to be applied. In a subsequent development, at the Lamont Geophysical Observatory

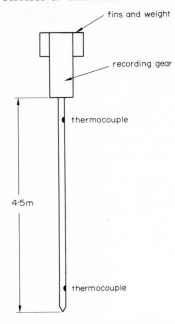

Figure 7.5 Bullard probe for sea floor heat flow

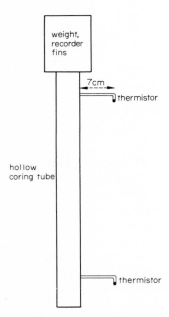

Figure 7.6 Lamont probe for sea floor heat flow

in the U.S.A., the thermometers were placed at the ends of fins projecting from the corer tube (figure 7.6). The fins were small enough not to heat the mud significantly as they were forced through it, while they were far enough from the tube for the heat generated by the penetration of the main tube to take some minutes to diffuse out to them. In this way the undisturbed temperature of the mud can be obtained almost immediately after the corer has been driven in. A further advantage of this second system is that the corer takes a sample of the actual mud in which the temperatures are measured.

The conductivity of the sediments at the sea bed depends greatly on the amount of water in them, and it is therefore important to take undisturbed samples from which the water content can be determined. The conductivity itself need not be measured, for it has been found that the reciprocal of the conductivity, the thermal resistivity, bears a linear relation to the water content (figure 7.7).

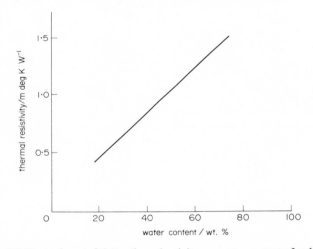

Figure 7.7 Dependence of thermal conductivity on water content of sediments

Many measurements of heat flow have now been made, the great majority of them at sea, and the results are summarized in the map of figure 7.8. Table 7.1 gives the mean values for a number of areas. The main conclusion to be drawn from these data is that the heat flow depends very little upon where it is measured,

Table 7.1   Representative values of heat flow

Area	Heat flow ($W\ m^{-2}$)
all continents	0·063
Atlantic Ocean	0·055
Pacific Ocean	0·073
Indian Ocean	0·060
world	0·060

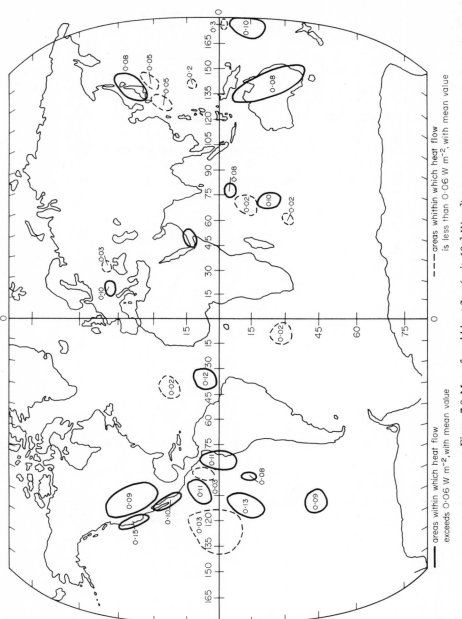

Figure 7.8 Map of world heat flow (unit, 10^{-2} W m^{-2})

——— areas within which heat flow
exceeds 0·06 W m^{-2}, with mean value

- - - areas whithin which heat flow
is less than 0·06 W m^{-2}, with mean value

a surprising conclusion in view of the higher concentration of radioactive elements in rocks of the continental crust. Apart from the general uniformity of the heat flow, there are a few small areas where the flow is appreciably higher, mainly on the mid-oceanic ridges.

7.3 Sources of heat within the Earth

The aim of the theoretical study of the heat flow from the Earth is to investigate the possible sources of heat within the Earth and the likely variation of tempera-ture within the Earth, for the electrical and mechanical properties of the material of the Earth are both strongly dependent on temperature, while ideas about the origin and constitution of the Earth and the other planets are closely connected with the nature of the source of the heat now flowing out from the Earth. If we consider just those parts of the Earth that are rigid, the flow of heat obeys the equation

$$\operatorname{div}(\kappa \operatorname{grad} T) = \rho C \frac{\partial T}{\partial t} - H \tag{7.4}$$

The symbols κ, ρ and C have the meaning given them in equation (7.1) while H is the rate of generation of heat per unit volume within the Earth. If the properties of the Earth are supposed to depend only on radius, the equation reads

$$\frac{1}{r^2} \frac{\partial}{\partial r}\left(\kappa r^2 \frac{\partial T}{\partial r}\right) = \rho C \frac{\partial T}{\partial t} - H(r, t) \tag{7.5}$$

The difficulty in obtaining realistic solutions of the equation of heat transfer within the Earth is that neither κ nor H is known directly at any great depth below the surface, and so values must be assigned on the basis of theoretical extra-polation from our knowledge of the behaviour of materials in the laboratory, or they must be based on inferences from other data about the Earth and the solar system. There are similar, and even more intractable, problems in trying to estimate the transport of heat by convection within the Earth. In this section, evidence for the sources of heat in the Earth is set out, the processes of heat transfer are considered in the next section, and lastly some indication is given of the possible thermal state of the interior. There are many uncertainties and gaps in our knowledge, some of which it may perhaps never be possible to fill, and the main aim in the remainder of this chapter must therefore be to indicate the arguments that can be used and to emphasize those results that do seem to be well established.

Consider first the original heat of the Earth, equal to its heat capacity multiplied by the temperature, which it had when it started to cool as an independent member of the solar system. It is not too difficult to estimate the heat capacity of the materials that probably constitute the Earth, but any estimate of the original temperature depends very much on how it is thought the Earth was formed. If the

Earth was formed as a hot molten body, then the temperature could be taken to be the melting temperature, for convection would rapidly reduce it to that value. If, on the other hand, the Earth originated by the accumulation of cold rock and dust, then the temperature at any radius would initially be determined by the balance between radiation from the surface and the rate at which energy was received from the Sun—probably not the same as at present. As the rock surface was covered with further material, the temperature would start to rise as radioactive heating became effective but the initial temperature to be taken at any radius would be the surface temperature at the time of formation. The initial conditions thus depend on the rate at which the Earth was formed for, if it formed fast relative to a characteristic time for radioactive heating, then the initial heat content would be the thermal capacity multiplied by the surface temperature. A third possibility is that the core formed first and that the mantle then accumulated around it; in that case the appropriate temperatures would seem to be the melting temperatures of the respective materials, for this possibility is related to the idea that, at the time of formation of the Earth, the Sun extended to the orbit of Mercury and the temperature of material in the neighbourhood of the Earth was close to the melting point of core material. There are thus two quite distinct possibilities, that the temperature of the Earth was low, say about 100 °K, when formed, or that it was high, 1500 °K or more. The difference between the two conditions corresponds to a heat content of some 8×10^{30} J, an amount that it would take about 10^4 My to dissipate at the present rate of outflow from the surface; the choice of a hot or cold initial state therefore has a very considerable effect on the subsequent behaviour.

The first source of heat to be considered within the Earth is the decay of radioactive nuclei. The rate of generation of heat by the isotopes at present found in the crust was given in table 6.1. The amounts of the radioactive isotopes in typical rocks are given in table 7.2 and they lead to the rates of heat generation listed there (see example 7.1). Suppose that the typical section of continental crust consists of a layer of perhaps 5 km of acidic composition, including sediments, lying upon 25 km of materials of intermediate composition. Using the figures in table 7.2, the heat generated in the continental crust is $4 \cdot 2 \times 10^{-2}$ W m^{-2} which is two-thirds of the observed flow of heat through the surface of the Earth. For the oceanic crust, take 5 km of water, 0·5 km of sediments of acidic composition, 2 km of rocks of basic composition and the remaining 22·5 km (oceanic crust plus upper mantle) of ultrabasic composition. The rate of heat generation is then $2 \cdot 5 \times 10^{-3}$ W m^{-2}, very much less than that from the continents (see example 7.2). In view of these figures it is difficult to understand why the heat flow from the continents is almost the same as that from the oceans. It seems likely that the radioactive content of the continental crust has been greatly overestimated, and there are direct observations that suggest that the high values usually quoted for granite are not typical. Studies of a batholith in the Rocky Mountains have shown that the uranium and thorium lie almost entirely in the upper skin of the rock mass and it may be that that is generally the case. The content of potassium does not vary throughout the mass of granite, and by itself produces heat at the rate of 10^{-10} W kg^{-1} or about one-tenth of the total radioactive heating in granite; it may be that the amount of granite in the crust has been overestimated, for the properties of the crust as a whole are not very like those of granite.

The rate of heat generation in the early stages of the life of the Earth was much greater than it is now. In the first place, it is a simple matter to calculate from the decay constants of uranium, potassium and thorium that the rate was about four times the present rate 4500 My years ago. In addition, any isotopes with shorter lives that could have existed in the Earth in its early history have now disappeared altogether. Such isotopes are aluminium-26, beryllium-10, chlorine-36, iron-60, and others, and their half-lives are of the order of 1 to 10 My. The Earth would have had to grow to almost its present size in a few million years for the heat generated by the short-lived isotopes to have contributed significantly to the

Table 7.2   Heat generation in typical rocks

Rock	Chemical content (parts per million)			Heat generation (W kg^{-1})
	U	Th	K	
granite	4·0	18	35000	94×10^{-11}
diorite	2·0	7	18000	42
basalt	0·8	3	8000	17
eclogite	0·04	0·2	1000	1·2
peridolite	0·01	0·06	10	0·25
dunite	0·001	0·004	10	0·02
chondritic meteorite	0·01	0·04	840	0·45

Note. In calculating the heat generation, the following abundancies are used:
^{238}U/U : 0·993
^{235}U/U : 0·007
^{40}K/K : $1·2 \times 10^{-4}$

heating of the Earth in its early stages, the only way in which their influence could still be felt. It seems that the short-lived isotopes have had no important effect on the temperature of the Earth.

Table 7.2 contains, as well as data for rocks, data for the chondritic meteorites, the average composition of which is sometimes thought to be a possible model for that of the Earth; the rate of heat production in these bodies is indeed very similar to that in the Earth as a whole (example 7.3).

A second source of heat within the Earth may be the release of gravitational energy. If the Earth were formed from particles of uniform composition, then at some time the core developed by the separation of the heavier material towards the centre. Gravitational energy would be released and would heat the Earth. If, however, the Earth were originally molten, and the core separated in the liquid phase or, if the core formed first and the mantle accumulated round it, then in either case gravitational heating would not affect the initial conditions, that the Earth was at its melting temperature, nor would it be effective in the subsequent history of the Earth.

The relative importance of different possible sources of heat depends very much on the hypothesis adopted for the origin of the Earth. If the Earth were assembled from cold particles, gravitational differentiation supplied a significant amount of heat and contributed to the heating up of the Earth from the cold state in which it was formed. If the Earth were formed in a hot state, either as a homogeneous body that then developed a core (now a somewhat outmoded idea) or as a hot core on which a mantle accumulated, then gravitational differentiation of the core would have made no contribution to the heating, but the original heat of the Earth would be significant. In either case, there is some difficulty in seeing how gravitational heating or original heat is to be reconciled with a present heat flow that could be accounted for entirely by radioactivity, although the interpretation to be put on the near equality between continental and oceanic heat flows may be that radio-active heating, which is greater immediately under the continents than under the oceans, is as a whole less important than surface measurements of radioactive content would imply.

7.4 Processes of heat transport

Heat is transported near the surface of the Earth by conduction at a rate that can be found from experiments in the laboratory, but we know enough of the processes of heat transport in crystals at high temperature and pressure to be sure that they are quite different at even moderate depths within the Earth.

In metals, heat is transported by the motions of electrons, but electrons play a very small part in the materials of the mantle of the Earth which, under the pressures and temperatures of the mantle, are ionic crystals that are semi-conductors having few free electrons. In such materials, one process by which heat transport takes place is that the motions of the crystal lattice depart slightly from simple harmonic vibrations in the presence of a gradient of temperature, that is when the amplitude of vibration, and therefore the degree of anharmonicity, vary with position while at the same time the speed of sound waves varies. Defects in the crystal reduce the conductivity, but the overall result is the same, that the conductivity is approximately inversely proportional to the absolute tempera-ture. On the other hand, the conductivity increases with pressure and it has been calculated that at depths less than about 500 km the lattice conductivity of the mantle will be less than at the surface, whereas at greater depths it will exceed the surface value.

Many silicate minerals are more or less transparent to infra-red radiation. In such conditions, heat is transported by infra-red radiation and experiments show that at high temperatures the thermal conductivity of a material such as olivine departs increasingly from the value (proportional to $1/T$) expected for lattice conduction (figure 7.9).

The effective thermal conductivity for heat transported by radiation is given by

$$\kappa_r = \frac{16}{3}\frac{n^2}{\eta}\sigma T^3$$

In this formula, σ is the Stefan–Boltzmann constant, n is the refractive index and η is the coefficient of absorption. The absorption coefficient of olivine, for example, has a minimum for radiation of wavelengths around 3 μm and so, if the temperature is high enough for the density of black body radiation in this band to be appreciable, radiation will play a significant part in the transport of heat, and that part will increase with temperature, through the factor T^3 in the formula for the conductivity, even though η increases as the temperature rises.

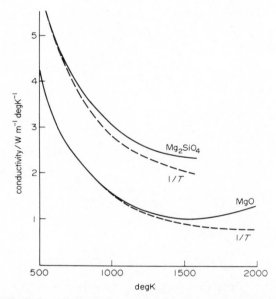

Figure 7.9 Variation of thermal conductivity of minerals with temperature

The materials which form the mantle are semiconductors at high temperature (see chapter 9) and the states of the electrons fall into two bands—valence bands in which the electrons are localized on particular ions and conduction bands in which the electrons are free to move through the crystal. The reduction of intensity of a beam of radiation as it travels through a crystal such as olivine is due to three processes: scattering from irregularities in the crystal ('grey' atmosphere), excitation of electrons from the valence to the conduction band, and excitation of ions.

Excitation of electrons from the valence to the conduction band gives a continuous spectrum of absorption since electrons can be excited by radiation of any energy greater than the band gap energy E_g (chapter 9). The corresponding absorption coefficient is

$$\frac{a}{v} \exp \left(-\frac{E_g}{kT} \right)$$

(a is a constant).

Now the electrical conductivity in a semiconductor follows the law

$$\sigma = \sigma_0 \exp \left(-\frac{E_g}{kT} \right) \quad \text{(chapter 9)}$$

(here σ is the electrical conductivity) and so it should be possible to calculate E_g from the variation of electrical conductivity in the mantle. However, absorption

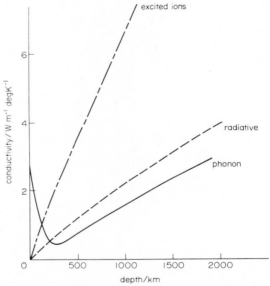

Figure 7.10 Variation of components of conductivity within the Earth

by the excitation of ions appears to be more important within the mantle than continuous absorption through transitions to the conduction band. Excitation of ions may take place if the energy of the photons is less than E_g; and transport of heat occurs when an excited ion transfers the excitation energy to an unexcited ion. Exciton conductivity increases rapidly with temperature and, as indicated by figure 7.10, appears to dominate other mechanisms in much of the mantle.

The third process of heat transport is by convection, in which heat is conveyed by the movement of hot material into a region of lower temperature. If a layer of liquid is heated from below, the hot liquid expands and the pressure of the surrounding liquid exerts on it a buoyancy pressure equal to $p\delta V$, where δV is the change of volume from thermal expansion. The buoyancy pressure is upwards and the liquid at the bottom of the layer would rise upwards through the layer were it not resisted by the viscous force exerted upon it by the surrounding liquid. The upward buoyancy force is proportional to

$$g\alpha\beta$$

where g is the value of gravity, α the coefficient of linear expansion and β the temperature gradient; the resistance is proportional to

$$\kappa v/\rho C_p$$

where κ is the thermal conductivity, ρ the density, C_p the specific heat at constant pressure and v the kinematic viscosity equal to the viscosity, η, divided by the density ρ.

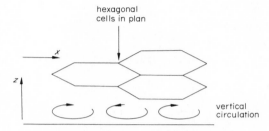

Figure 7.11 Bénard cell in convection

In a layer of thickness t, the balance between the upwards force and the drag is expressed by the ratio

$$R = \frac{g\alpha\beta\rho C_p}{\kappa v}t^4$$

a non-dimensional quantity known as the *Rayleigh number*, after Lord Rayleigh who first investigated convection in a layer of liquid heated from below. Rayleigh showed that convection would occur if the number had a value of about 10^3 and he also showed that the first movement would be in a simple cellular pattern (figure 7.11) in which the horizontal dimension of a cell is about π times the thickness of the liquid layer.

If the fluid is incompressible, β is the actual temperature gradient but, if the fluid is compressible, it will be stable provided the gradient does not exceed the adiabatic value, $g\alpha T/C_p$. According to the data given in example 7.4, the adiabatic gradient in the liquid core of the Earth must be close to 0.15 degK km.$^{-1}$

Convection can occur not only in a horizontal layer but also in a spherical shell or in a sphere of liquid with a source of heat in the interior. The Rayleigh number for the onset of convection differs from that for the simple horizontal layer but the order of magnitude is the same.

The simplest patterns of the movement of a fluid occur when the Rayleigh number only just exceeds the critical value and the speed of the motion is very small. As the supply of heat is increased and the movement is speeded up, other, more complex, patterns develop. Cn the other hand, additional forces may prevent the simplest pattern developing, and then convection will not start until the rate of supply of heat is great enough for some other pattern to come into being. Within the Earth there are two important additional forces, those arising from the spin of the Earth and those arising from the magnetic field within it and, on account of these forces, it may be expected that convection will not start until the Rayleigh number exceeds the value for the onset of convection in a simple horizontal layer.

In the core

$$g = 5 \text{ m s}^{-1}$$

$$\alpha \sim 5 \times 10^{-5} \text{ degK}^{-1}$$

$$t \sim 3000 \text{ km}$$

$$\rho \sim 12\,000 \text{ kg m}^{-3}$$

$$C_p \sim 500 \text{ Jkg}^{-1} \text{ degK}^{-1}$$

$$v \sim 5 \times 10^{-7} \text{ m}^2 \text{ s}^{-1}$$

$$\kappa \sim 3 \text{ Wm}^{-1} \text{ degK}^{-1}$$

and the Rayleigh number would be about 10^{32} if the actual gradient were to exceed the adiabatic gradient by only $0 \cdot 1$ degK km^{-1}. Convection therefore appears certain to occur in the core and it may be expected to dominate other modes of transport with the actual gradient very close to $0 \cdot 15$ degK km^{-1}.

The values of the properties of the iron core have been taken from values for molten iron in the laboratory, but R is so great that even changes by an order of magnitude will scarcely affect the argument. It is very different with the mantle. Convection in the mantle has been discussed by many authors (see Knopoff, 1967) and two conclusions seem important. In the first place, it may well be that convection cannot cross the boundary between the upper and lower mantle, and so convection in the upper mantle and convection in the lower mantle have to be discussed separately with corresponding separate values of the length t and the viscosity. Secondly, the evidence for the viscosity of the mantle comes mainly from the phenomenon of uplift following the melting of the ice caps in Scandinavia and over Lake Bonneville (chapter 5), and this evidence suggests a value of about 10^{25} m^2 s^{-1} for the kinematic viscosity of the upper mantle. A higher value must, however, apply in the lower mantle, partly to support the flattening of the Earth which is greater than the hydrostatic value (chapter 2) and partly to satisfy the observations in Scandinavia which show different rates of recovery in different areas. The viscosity of the lower mantle may be as high as 10^{30} m^2 s^{-1}. Convection would then not be possible in the lower mantle but might occur in the upper mantle.

In estimating Rayleigh numbers, the material of the mantle has been treated as though it were a viscous fluid, yet we know (chapter 5) that that is an unwarranted simplification, for in the first place the mantle transmits shear waves and it behaves as though it had a slight strength in shear. Nonetheless, it is supposed that, over a long enough time, it may be treated as a liquid and that convection will occur. This is a supposition and assumes that some very difficult questions have answers. In the first place, the viscosity and the shear strength certainly depend on temperature—how, in consequence is the onset of convection affected? Secondly, the sources of heat, the radioactive elements of the mantle, are distributed through the mantle and not concentrated at the base—it is not clear how this circumstance affects the theory of convection. Finally, if the mantle has a finite strength, even though a small one, over an arbitrarily long time, can convection still take place or, if so, will it resemble at all closely the convection we observe in water or the atmosphere?

If the conductivity, both lattice and radiative, of the upper mantle is not too great, then the temperature may be expected to attain a value at which the rate of creep is great enough for motion to occur under the buoyancy forces. This is quite a different condition from that for an ordinary fluid where the variation of viscosity with temperature may be neglected to a first approximation and the main effect of increasing temperature is to increase the buoyancy forces. In the likely circumstances within the mantle, the main effect of increasing temperature would be to increase the rate of creep. Again, because the creep rate depends strongly on temperature, the patterns of fluid motion are most unlikely to be similar to those in a fluid with a constant viscosity. These remarks will indicate the very severe problems that lie in the way of studying the possible convective transport of heat within the Earth.

The upshot of this discussion seems to be that convection dominates in the core, where the gradient must be close to the adiabatic gradient, that convection is unlikely in the lower mantle, and that some form of convection may occur in the upper mantle but it is not yet possible to calculate the temperature distribution there.

7.5 The internal temperature of the Earth

It may seem at first sight that it is an almost hopeless task to estimate the internal temperature of the Earth in view of the uncertainties about both the sources of heat and the processes of heat transfer within the Earth. Something can, however, be made of the situation. The temperature within the core must be close to the adiabatic gradient, of the order of 0.15 degK km^{-1}, and the temperature at the boundary of the inner core may be the melting point under the pressure at that radius. From shock wave data (chapter 5) that temperature is estimated to be about 4000 °K. The temperature at the outer boundary of the core might then be in the neighbourhood of 3600 °K.

Even given the temperatures at the top and bottom of the mantle, there seem to be many obstacles in the way of interpolating between them, but the problem may be to some extent simplified on account of the fact that the rate of transfer of heat by radiation depends on the cube of the temperature:

$$\kappa = \kappa_0 + \kappa_1 T^3$$

where

$$\kappa_1 = \frac{16}{3} \frac{n^2 \sigma}{\eta}$$

If radiation transfer dominates, κ_0 may be neglected and the equation of heat transfer reads:

$$\frac{1}{r^2} \frac{\partial}{\partial r} \left(\kappa_1 T^3 r^2 \frac{\partial T}{\partial r} \right) = \rho C \frac{\partial T}{\partial t}$$

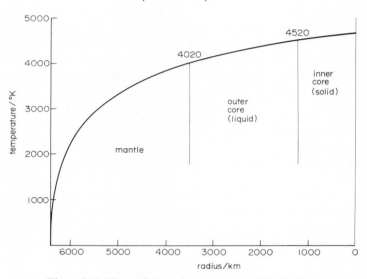

Figure 7.12 The variation of temperature within the Earth

Further suppose that κ_1 is great enough that the Earth is at present in thermal equilibrium and $\partial T/\partial t = 0$.

Then

$$\frac{\partial}{\partial r} \left(\kappa_1 T^3 r^2 \frac{\partial T}{\partial r} \right) = 0$$

if there are no sources of heat distributed in the mantle.

If there is a uniformly distributed source of heat H per unit volume,

$$\frac{\partial}{\partial r} \left(\kappa_1 T^3 r^2 \frac{\partial T}{\partial r} \right) = -r^2 H$$

As shown in example 7.5, these equations have solutions which depend on $r^{1/4}$ and $r^{1/2}$ respectively and so vary relatively slowly with radius. The radiative transfer of heat increases rapidly with temperature, and therefore with the heat supplied and, for that reason, the temperature established in the mantle is rather insensitive to the rate of supply of heat.

Many numerical studies of the distribution of temperature within the Earth have been made (see Lubimova, 1967), some of which show the stabilizing effect of radiative transfer of heat. If the temperature at the boundary between the core and mantle is fixed, as suggested at the beginning of this section, at 3600 °K, and if the temperature at the surface is 300 °K then the temperature within the mantle must follow a course very like that plotted in figure 7.12.

Examples for chapter 7

7.1 Using the data of table 6.1 and the abundance ratios and chemical contents given in table 7.2, calculate the heat generation in W kg^{-1} of granite, basalt and dunite and compare your values with those given in table 7.2.

7.2 Suppose that the composition of continental and oceanic crustal columns are as follows:

Continents		Oceans	
granite, 2700 kg m^{-3},	5 km	water 1030 kg m^{-3},	5 km
intermediate 2800 kg m^{-3},	25 km	sediment 2400 kg m^{-3},	0·5 km
		basaltic 3000 kg m^{-3},	2 km
		mantle 3300 kg m^{-3},	22·5 km
		(periodite)	
Totals	30 km		30 km

Calculate the rates of heat generation per square metre of continental and oceanic sections, using the data in table 7.2.

7.3 Given that the rate of outflow of heat from the Earth is 0·06 W m^{-2}, calculate the rate of loss of heat per kg averaged throughout the Earth, and compare it with the rate of radioactive generation of heat in a chondritic meteorite. (Surface area of Earth, $5·1 \times 10^{14}$ m^2, mass of Earth $6·0 \times 10^{24}$ kg.)

7.4 Calculate the adiabatic gradient in the core of the Earth from the following data:

g	10 m s^{-2}
α (coefficient of linear expansion)	10^{-5} degK^{-1}
T	4000 °K
C_p	500 J kg^{-1}

7.5 Solve the equation of conduction for radiative transfer under steady state conditions ($\partial T/\partial t = 0$) in an infinite slab with the temperature a function only of depth (z) in the slab: (a) for no distributed heat sources; (b) for uniformly distributed heat sources.

The equation is

$$\frac{d}{dz}\left(\kappa T^3 \frac{dT}{dz}\right) = 0 \qquad\qquad\text{(a)}$$

$$= -H \qquad\qquad\text{(b)}$$

(Assume $T = T_0 z^p$.)

7.6 Solve the equation for conduction of heat under steady state conditions ($\partial T/\partial t = 0$) in a spherical shell with purely radial variation of temperature (a) for no distributed heat sources; (b) for uniformly distributed heat sources. The equation is

$$\frac{\partial}{\partial r}\left(\kappa r^2 T^3 \frac{\partial T}{\partial r}\right) = 0 \qquad\qquad\text{(a)}$$

$$= -H \qquad\qquad\text{(b)}$$

(Assume $T = T_0 r^p$.)

Further reading for chapter 7

Bullard, E. C., 1939, *Proc. Roy. Soc.*, Ser. *A*, **173**, 474–502.

—— 1966, 'The flow of heat through the floor of the ocean', in *The Seas*, Vol. III (Ed. M. N. Hill) (New York, London, Sydney: Wiley–Interscience), xvi + 963 pp.

von Herzen, R. P., 1967, 'Surface heat flow and some implications', in *The Earth's Mantle* (Ed. T. F. Gaskell) (London, New York: Academic Press), xiv + 509 pp.

Knopoff, L., 1967, 'Thermal convection in the Earth's mantle', in *The Earth's Mantle* (Ed. T. F. Gaskell) (London, New York: Academic Press), xiv + 509 pp.

Lubimova, E. A., 1967, 'Theory of thermal state of Earth's mantle', in *The Earth's Mantle* (Ed. T. F. Gaskell) (London, New York: Academic Press), xiv + 509 pp.

CHAPTER EIGHT

The permanent magnetization of rocks

'The palest ink is better than the lost memory.'
Chinese proverb

8.1 Introduction

Ever since the time of William Gilbert in the reign of Queen Elizabeth the First, it has been known that the Earth has a magnetic field which is very close to that of a dipole at the centre. Many rocks contain ferromagnetic minerals and so it might be expected that they would acquire a small magnetization parallel to the field in which they find themselves. Indeed, many rocks that were originally hot, in particular lavas that have flowed out of volcanoes, have acquired a permanent magnetization when they cooled from a high temperature, at which the minerals they contain are paramagnetic, to a lower temperature at which the minerals are ferromagnetic. The temperature at which the change takes place is called the Curie temperature and is in the neighbourhood of 500 °C, well below the melting point of lavas, so that it is quite common for lavas to acquire a permanent magnetization in the direction the local field had somewhat after they solidified. The direction of the dipole at the centre of the Earth is close to the polar axis, but does not coincide with it and at the present time, for example, the north magnetic pole lies in the region of Hudson's Bay in Canada. The direction is not constant but undergoes a slow change known as the secular variation and, correspondingly, the direction of the magnetic field at a particular point on the surface of the Earth changes slowly. By measuring the direction of the permanent magnetization of lavas of which the dates are known from historical evidence, it is possible to trace out the secular variation of the magnetic field in past times. Evidence from lavas may be supplemented from measurements of the magnetization acquired by clay pots when they cooled from firing and conversely, if the secular variation has been mapped out, the date of the firing of a clay pot can be estimated, provided the article is found in the position in which it was fired or, better, if the magnetization of the furnace hearth (which is less likely to have moved) is studied.

Suppose, first, that the central dipole is directed along the polar axis, as shown in figure 8.1. Then the potential due to the dipole at a point with coordinates (r, θ, ϕ) is

$$V = \frac{S}{r^2} \cos \theta \tag{8.1}$$

where S is the strength of the dipole.

The radial component of the field is $-\partial V/\partial r$, that is

$$B_r = \frac{2S}{r^3}\cos\theta \qquad (8.2)$$

Similarly the horizontal component along the meridian, B_θ, is

$$-\frac{1}{r}\frac{\partial V}{\partial \theta} = \frac{S}{r^3}\sin\theta \qquad (8.3)$$

while the horizontal component in the east–west direction perpendicular to the meridian is zero since the potential is independent of the longitude. At the

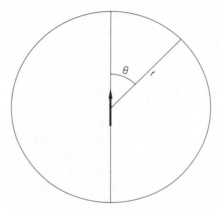

Figure 8.1 Axial dipole

surface of the Earth, r is equal to the mean surface radius, a,

$$\left. \begin{array}{l} B_r = \dfrac{2S}{a^3}\cos\theta \\[2mm] B_\theta = \dfrac{S}{a^3}\sin\theta \end{array} \right\} \qquad (8.4)$$

and the resultant field B is equal to

$$\frac{S}{a^3}(3\cos^2\theta + 1)^{1/2}$$

and is inclined at an angle I to the horizontal, such that

$$\tan I = 2\cot\theta \qquad (8.5)$$

Thus at the poles the inclination is $\frac{1}{2}\pi$, while at the equator it is zero.

Supposing, then that the Earth's field is that of a dipole at the centre, the angle between the latitude of the site of a pottery kiln or a lava flow and the direction of the axis of the dipole can be found from a measurement of the inclination of the permanent magnetization of the pots or the lava.

The more general situation, where the axis of the dipole is in some arbitrary direction, is indicated in figure 8.2. If ψ is the angle between the direction of the dipole and the direction of the radius vector of a point at the surface of the Earth, then the potential at the latter point is

$$\frac{S}{r^2}\cos\psi$$

Now

$$\cos\psi = \cos\theta_D\cos\theta + \sin\theta_D\sin\theta\cos(\phi_D - \phi)$$

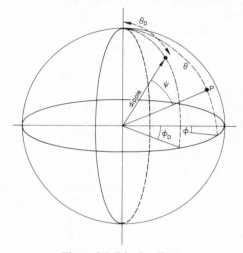

Figure 8.2 Dipole off axis

where (θ_D, ϕ_D)) are the coordinates of the dipole vector, and so

$$
\left.
\begin{aligned}
B_r &= -\frac{\partial V}{\partial r} = \frac{2S}{a^3}\{\cos\theta_D\cos\theta + \sin\theta_D\sin\theta\cos(\phi_D - \phi)\} \\[2mm]
B_\theta &= -\frac{\partial V}{r\partial\theta} = \frac{S}{a^3}\{\cos\theta_D\sin\theta - \sin\theta_D\cos\theta\cos(\phi_D - \phi)\} \\[2mm]
B_\phi &= -\frac{\partial V}{r\sin\theta\,\partial\phi} = -\frac{S}{a^3}\sin\theta_D\sin(\phi_D - \phi)
\end{aligned}
\right\} \quad (8.6)
$$

As in the simple case, the direction of the central dipole can be found from measurements of the magnetizations of lavas or pottery.

Rocks in the more distant past may also have acquired a permanent magnetization in the field of the Earth at some past time, and it should be possible to reconstruct the past field from observations of the present magnetization. Indeed it can, but there are serious difficulties in the way. In the first place, rocks are frequently found at present in neither the place nor the attitude in which they were formed; in particular, rocks which now occur or did in the past occur in mountain ranges may be found to be vertical instead of horizontal or may even have turned right over. There must be convincing evidence that rocks to be used for the study of the past magnetic field lie in the place and in the attitude in which they were formed. Secondly, the magnetization now possessed by a rock may not all have been acquired at a single instant of time, as a pot or a layer of lava cooled through its Curie temperature; materials can also become magnetized just by sitting in a steady field for a long time. Tests must therefore be made to show that the magnetization does represent the field at some particular epoch. Equally, it must be shown that the magnetization originally acquired has not decayed away, for then it might be that the direction now found in the rock is not that originally induced. A further complication (or perhaps simplification) is that rocks only a few centimetres thick may have been formed over a period of time covering many cycles of the secular variation, and thus the measured directions may cover a range around the mean direction of the field, the secular variation being averaged out.

While recent rocks are magnetized by cooling in the field of the Earth, rocks which have accumulated over a period of time may have been magnetized in other ways as well. Sandstones that contain red oxides of iron in the cement between grains of sand are often magnetized, the iron oxide having acquired its permanent magnetization when it was precipitated from solution. Such red sandstones are useful for studies of the magnetic field in the past because they were often formed in regions where mountains have not developed and so are still relatively undisturbed even though sometimes of great age. Rocks may also acquire magnetization when they are deposited in water because grains of magnetic material orient themselves in the direction of the field as they fall through the water. However, as the sediments lose water and become more compact, the orientation of the minerals in the deposit will usually change. The importance of this type of magnetization is that it is the only one that is effective in the accessible rocks of the sea floor, those that were deposited fairly recently and that are still uncompacted.

8.2 Units of magnetic measurements

Magnetometers measure either the force upon a permanent magnet or the flux linking a coil; thus they measure the induction, B, the unit of which in the S.I. system is the tesla, equal to 10^4 gauss. The Earth's field is about 50 μT, and variations of the field are commonly measured in the unit called the gamma (γ) which is 10^{-5} G or 1 nT (nanotesla).

From the relation

$$B = -\text{grad } V \quad \text{(see equation (8.2))}$$

it follows that the units of dipole moment, S, are tesla metre^3 and hence that of intensity of magnetization, or magnetic moment per unit volume, is the tesla.

Magnetic flux has the units tesla metre^2 and, since the e.m.f. induced in a coil is equal to the rate of change of flux, it follows that

$$1 \text{ volt} = 1 \text{ tesla metre}^2 \text{ second}^{-1}$$

or

$$1 \text{ tesla} = 1 \text{ volt second metre}^{-2}$$

The unit of dipole moment is then the volt second metre, and 1 e.m.u. of dipole moment is equal to 10^{-10} T m^3.

8.3 The magnetic properties of rocks

The magnetic properties of rocks are due to the small amounts of ferrimagnetic minerals that they contain and, although the minerals themselves may be quite highly magnetized, they occur in such small quantities in most rocks that the magnetization of the rocks is very weak. The magnetization of the ferrimagnetic minerals comes from the magnetic moments of the ions of iron, Fe^{2+} and Fe^{3+}, and the minerals fall into two distinct classes according to the way in which the Fe^{2+} and Fe^{3+} ions are arranged in the crystal lattices. The first group is strongly ferrimagnetic, with spontaneous magnetizations attaining 0·05 T, comparable, that is, with the spontaneous magnetizations of the ferromagnetic metals. The minerals in this group have a cubic structure and consist of solid solutions of the two oxides of iron, magnetite (Fe_3O_4) and maghemite (Fe_2O_3) with titanium dioxide (TiO_2) comprising the titanomagnetites. In magnetite the Fe^{2+} and Fe^{3+} ions are arranged on different types of site, A and B, within the cubic lattice, and antiferromagnetic coupling between the two sites causes the moments of the respective ions to be aligned in opposite directions. But there are twice as many B sites as A sites and so there is a strong net spontaneous magnetization (figure 8.3).

The other group of minerals has rhombohedral structure and comprises haematite (Fe_2O_3) and its solid solutions with ilmenite ($FeTiO_3$). Pyrrhotite, FeS_x, where x lies between 1 and 2·1, is similar to haematite. In these crystals, the Fe^{3+} ions lie on intersecting lattices and the interaction between them causes them to lie almost anti-parallel but not quite (figure 8.4). Thus the net moment is at right angles to the mean direction of the ionic moments and is very much less, so that the spontaneous magnetization is of the order of 2×10^{-4} T.

As with ferromagnetic metals, a crystal of either of the foregoing ferrimagnetic minerals has least magnetic energy in zero field if it is divided into a number of domains with the magnetic moments in opposite directions. The size of a domain depends on the spontaneous magnetization and is about 0·03 μm for magnetite but more like 0·15 cm for haematite. The magnetite grains normally found in rocks are large enough to contain a number of domains but the grains of haematite are commonly single domains.

When an external magnetic field is applied to a magnetic mineral, the total energy will be reduced if those domains that are magnetized in the opposite direction to the applied field become remagnetized parallel to the applied field. The chance of remagnetization taking place is proportional to

$$e^{-E/kT}$$

where E is the difference between the energies corresponding to the two directions of magnetization. At a temperature of about 300 °K, it would take some 10^{17} y for the magnetization of magnetite to decay spontaneously and it is for this reason that the ancient magnetic field can be inferred from the magnetization of rocks. If, however, the mineral is heated, two things happen, thermal fluctuations increase and help the energy barrier to be overcome, while the barrier itself,

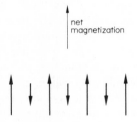

Figure 8.3 Ferrimagnet—magnetite type

Figure 8.4 Ferrimagnet—haematite type

which is proportional to the square of the spontaneous magnetization, decreases to vanish at the Curie temperature. Thus, above the Curie temperature, the mineral has no spontaneous magnetization and is not ferrimagnetic but, as it cools through the Curie temperature, it becomes magnetized while the energy barrier is still low enough for the domains to be aligned in an external field. In the presence of an external field, the mineral as a whole will acquire a magnetic moment, but then, as it cools further, the energy required to re-align the domains increases while the thermal fluctuations decrease and so below the so-called blocking temperature the moment induced by the external field near the Curie temperature is frozen into the mineral and can only be altered by an external field of intensity corresponding to the internal energy barrier to the movement of domains. That intensity is the coercive force of the mineral and is some $2·4 \times 10^4$ A m^{-1} for magnetite and more than $1·5 \times 10^5$ A m^{-1} for haematite. Magnetization acquired by cooling through the Curie temperature in a magnetic field is known as *thermoremanent magnetization* (T.R.M.) and is the magnetization of igneous rocks.

Sedimentary rocks do not acquire their magnetization by cooling because they were never hot (except in the special case of sediments that were at some time heated by contact with igneous rocks) but chemical process can also bring about magnetization. The haematite of red rocks is formed by the oxidation of iron minerals such as magnetite, and at first, when a grain is small, it can easily become magnetized by an external field in any direction. However, as it grows larger, the energy required to remagnetize it increases (it will be a single domain) and eventually it may grow to such a size that the field necessary to change its moment is much greater than the Earth's field. Magnetization so acquired is called *chemical remanent magnetization* (C.R.M.).

Ideally, a rock would be magnetized in the Earth's field at some unique time and would retain that magnetization thereafter, but in fact rocks can change their magnetization. The energy barrier to movement of domain boundaries is not an absolute one and there is a finite, though very small, chance of domains moving at normal atmospheric temperatures. Again, not all domain walls have the same energy barrier, so that, while some demains retain the form imposed on them at the formation of the rock, others may move in the ambient field. Rocks therefore usually display a resultant magnetization that is the vector sum of the moment initially induced at the formation of the rock together with the moment subsequently acquired in course of time. Fortunately, because the latter was easily acquired, it can also be easily removed by heating the rock or subjecting it to an alternating magnetic field, whereas the other component is far more resistant to demagnetization. Demagnetization tests to analyse the magnetization into primary and secondary components are an important part of the study of the magnetization of rocks.

The ferrimagnetic minerals in rocks show magnetic hysteresis in the same way as do ferromagnetic metals; an example for a typical igneous rock is shown in figure 8.5. If such a rock is subject to a varying magnetic field so that it is taken round an hysteresis cycle, then, when the external field returns to zero, the rock will remain magnetized with an intensity called the *isothermal remanent magnetization*; it is induced naturally if lightning strikes the ground, for the current of the lightning sets up a varying magnetic field which leaves the rocks in the neighbourhood magnetized after it has fallen to zero.

Rocks can also be magnetized through the alignment of magnetic grains. Consider a rock accumulating as grains of sediment settle in water. Non-magnetic grains falling through calm water align themselves so that the resistance to movement through the water is least but, if the grains are magnetized, then the alignment will be determined by a balance between the couple exerted by the Earth's magnetic field and the hydrodynamic couple. The sediment will become more compact as water is pressed out of it and grains that settled at an angle to the vertical will rotate to become more nearly horizontal. In general, therefore, rocks that have become magnetized through settling in water will show a declination close to the direction of the field in which they were formed, but the inclination will be less. Magnetization acquired through settling is known as *detrital remanent magnetization* and is found especially in the clays that settled in calm glacial lakes (varve clays) and in some sediments of the deep oceans.

In some very special circumstances, rocks can become magnetized in the opposite direction to the magnetizing field. Two conditions are necessary—the rocks should contain two types of mineral with different Curie temperatures, and they should show different magnetic properties in different directions. It may then happen that, as the rock cools through the successive Curie temperatures, the net field exerted on one of the minerals is in the opposite direction to

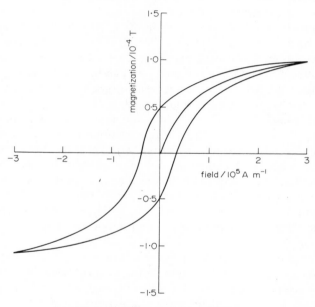

Figure 8.5 Hysteresis loop for lava

the Earth's field and the final magnetization at ambient temperature is also opposed to the Earth's field. Such mixtures of minerals have been made up in the laboratory and have been found to acquire reversed magnetization and, very infrequently, naturally occurring rocks have been found to show self-reversed properties. The possibility of self-reversed magnetization is important because of the observation (section 8.5) that reversed magnetization is quite commonly found to occur in nature; to understand the magnetic field of the Earth it is necessary to decide whether such reversals represent reversals of the magnetic field of the Earth or whether they arise from self-reversed magnetization in a field of the same direction as at present.

The susceptibilities of the magnetic minerals found in rocks are similar to those of ferromagnetic metals, but, because the minerals are found in very small quantities, the susceptibilities of rocks are very low, comparable with those of paramagnetic substances. The susceptibility varies with temperature in the same way as that of the minerals and, if more than one mineral is present, the susceptibilities varying with temperature in different ways, then an analysis of the overall variation shown by the rock will often enable the minerals to be identified. Similarly, the saturation magnetization varies with temperature in a characteristic manner for individual minerals so that, again, a study of the variation of the saturation magnetization of a rock as the temperature is varied may enable the constituent magnetic minerals to be identified.

The properties of some magnetic minerals and rocks are summarized in table 8.1.

Table 8.1   Magnetic properties of rocks and minerals

	Minerals			Rocks	
	Magnetite Fe_3O_4	Maghemite γ-Fe_2O_3	Haematite α-Fe_2O_3	Basalt	Red sandstone
Initial susceptibility	10^{-2} to 1		2×10^{-5}	1 to 8×10^{-3}	10^{-5} to 10^{-6}
Saturation magnetization (T)	$4 \cdot 5 \times 10^{-2}$	4×10^{-2}	0·5 to $2 \cdot 5 \times 10^{-4}$		
Coercive force (A m^{-1})	$2 \cdot 5 \times 10^4$		$1 \cdot 5 \times 10^5$	4×10^4	2×10^5
Curie point (°C)	578	675	700		
Typical remanent magnetization					
Thermal (T)	10^{-5}		10^{-3}	5×10^{-7}	
Chemical (T)					10^{-9}

Note. The magnetization of rocks and minerals is very variable.
Units

(1) Intensity of magnetization may be expressed in magnetic moment per unit volume or per unit mass. Here it is given per unit volume and therefore in terms of the tesla. In the alternative form, obtained by dividing by the density, the units would be T m^3 kg^{-1}. The c.g.s. units often employed are

$$\text{(a) e.m.u. cm}^{-3} = 10^{-4}\text{ T}$$
$$\text{(b) e.m.u. g}^{-1}\quad = 10^{-7}\text{ T m}^3\text{ kg}^{-1}$$

(2) The unit of coercive force (magnetic field strength) per metre (Am^{-1}) equal to $4\pi\ 10^{-3}$ oersted.

8.4 The measurement of the magnetic properties of rocks

The measurements usually made upon rocks comprise the determination of the intensity and direction of the remanent magnetization and the way in which they vary with temperature or in an alternating demagnetizing field, and the variation of susceptibility and saturation magnetization with temperature. The first group of measurements enables the primary remanent magnetization to be distinguished from that acquired subsequently in the history of the rock, as indicated in section

8.3, while the second group may allow the constituent minerals to be identified. The instruments that are used in such investigations are very sensitive magnetometers to measure the weak magnetization of rocks, furnaces to enable the properties to be measured at different temperatures, and systems of coils with which the rocks can be subjected to demagnetizing fields.

Two types of magnetometer are generally in use, the astatic magnetometer and the spinner magnetometer. The astatic magnetometer (figure 8.6) consists of two strongly magnetized permanent magnets rigidly connected together and suspended by a fibre that exerts a weak torsional couple upon the pair of magnets. If the moments μ of the magnets are equal and are aligned in opposite directions

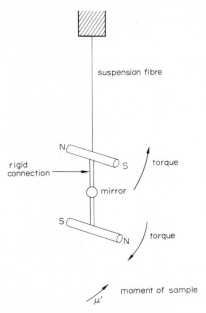

Figure 8.6 Astatic magnetometer

so that the net moment is zero, then in a uniform field the couple, $\mu \wedge B$, exerted upon one of the magnets will be equal and opposite to that exerted upon the other and so the net magnetic couple upon the pair will be zero. If, however, the field is not uniform, but has the value B_1 at one magnet and B_2 at the other, the net couple will be

$$\mu \wedge B_1 - \mu \wedge B_2$$

and the magnets will rotate until this net magnetic couple is balanced by the couple $\tau\theta$ exerted by the suspension fibre. The angle of twist of the magnet pair is then given by

$$\theta = (\mu \wedge B_1 - \mu \wedge B_2)/\tau.$$

Suppose the magnets to be separated by a distance l.

If a rock having a magnetic moment μ' is placed below the magnet pair, so that it is at a distance d from the lower magnet, the magnitude of the field it produces at the latter will be

$$\frac{\mu'}{d^3}$$

and that of the field it produces at the upper magnetic will be

$$\frac{\mu'}{(d+l)^3}$$

Thus the maximum couple that can be exerted on the magnetic pair will be

$$\mu\mu'\left\{\frac{1}{d^3}-\frac{1}{(d+l)^3}\right\}$$

from which, if μ, l and d are known, μ' may be calculated.

The deflection of the magnets is measured by observing the deflection of a beam of light reflected from a mirror attached to them. The magnets are made of commercial permanent magnet material having high remanence and high coercive force. Pieces of material are ground to size and pairs are chosen to have as nearly equal moments as possible. The stronger magnet of a pair is then demagnetized carefully until the two moments are very close and the final adjustment is made by attaching small trimming magnets to the main magnets. The period of oscillation of the suspended system is given by

$$2\pi\left(\frac{I}{\tau}\right)^{1/2}$$

where I is the moment of inertia of the two magnets. The deflection is proportional to μ/τ, but has to be read in the presence of the Brownian movement of the suspended system. Blackett showed that, if the period is given, then the best ratio of deflection to Brownian movement is obtained when $\mu/I^{1/2}$ is a maximum. However, the more practical consideration is that the period should be as short as possible consistent with high sensitivity, so that as many specimens as possible should be measured in a given time. In those circumstances, Creer showed that μ/I should be as large as possible, a condition that leads to small light magnets. Typical dimensions are magnets $6\,\mathrm{mm} \times 2\,\mathrm{mm}$ separated by 5 to 10 cm and having moments of about $10^{-9}\,\mathrm{T\,m^3}$.

The sensitivity of the magnetometer may be calibrated by observing the deflection when a known current is passed through a small coil placed in the neighbourhood of the magnets. The sensitivity of an astatic magnetometer is limited by the uniformity of the field in which it is placed and by vibrations of the ground that set the suspended system into oscillation. If the ambient field did not vary with time, the fact that it was not uniform would not matter, for it would just produce a steady deflection of the magnets, but a non-uniform field that varies with time produces a changing deflection that limits the accuracy of measurements. Many man-made sources produce non-uniform fields if they are relatively close to the magnetometer, and, in addition, ferromagnetic material in the neighbourhood of the instrument will give rise to non-uniform fields that will vary with time as the main field of the Earth varies. Thus it is important to make the measurements in a building free from ferromagnetic material and far from electrical machinery if the most delicate measurements are to be made. Similarly, the laboratory should be placed on ground as free as possible from vibration and again, therefore, far from machinery and traffic.

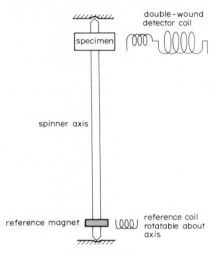

Figure 8.7 Spinner magnetometer

If sedimentary rocks are to be measured, the whole magnetometer may have to be put in a field-free region for the properties of these weakly magnetized rocks are affected to some extent even by the ambient field. The magnetometer is then placed inside a system of three pairs of Helmholtz coils with which the total field at the specimen and at the magnets can be reduced to zero. Non-uniform components of the ambient field are not annulled and the precautions described in the previous paragraph must still be taken.

Astatic magnetometers can detect fields of the order of 10^{-14} T, corresponding to a rock 1 cm³ cube, having a magnetization of 10^{-11} T at a distance of 10 cm.

The principle of the spinner magnetometer is shown in figure 8.7. Consider a dipole in the neighbourhood of a coil. If the dipole is rotated about its axis, the field at the coil will always be the same and no voltage will be induced in the coil. Suppose, on the other hand, that the dipole is rotated about an axis perpendicular to its own. If the axis of rotation is parallel to the axis of the coil, the field at the coil will again be unchanged and no voltage will be induced but, if the axis of rotation is perpendicular to the axis of the coil, the field at the coil will vary and a voltage will be induced, the maximum value being attained when the component of the field parallel to the axis of the coil is changing most rapidly, that is when the dipole is perpendicular to the axis of the coil. Consider then, a piece of rock being spun about an axis perpendicular to the axis of a pick-up coil. No voltage will be induced by the component of the magnetic moment parallel to the axis of rotation, but the component perpendicular to the axis will induce a voltage which will attain its maximum amplitude when the component is perpendicular to the axis of the coil, or, in other words, the voltage will pass through zero when the component is parallel to the axis of the coil. The voltage induced in the coil is equal to the rate of change of magnetic flux, that is to

$$n\dot{\boldsymbol{B}}.\boldsymbol{A}$$

where n is the number of turns of the pick-up coil, $\boldsymbol{A}$ its area (considered as a vector) and $\boldsymbol{B}$ the flux.

If the magnetic moment $\boldsymbol{\mu}$ is inclined at an angle θ to the axis of the coil, the field at the coil is

$$B = \frac{2\mu \cos \theta}{r^3}$$

and

$$\dot{B} = -\frac{2\mu \sin \theta}{r^3}\omega$$

where ω is the spin angular velocity.

Thus

$$V = -\frac{2nA\mu\omega}{r^3}\sin \theta$$

The magnetic moment may be calculated from the magnitude of the induced voltage given the dimensions of the coil, the distance of the specimen and the speed of rotation, and the direction of the moment can be found from the direction of the specimen when the induced voltage passes through zero. If, now, the specimen is rotated so that the moment so found is parallel to the axis of rotation, and the experiment repeated, the component of the moment perpendicular to the first will be found, and thus the vector moment of the rock will be completely specified. The volume of a specimen of rock is commonly 5 to 10 cm³ and the sample is spun at up to 500 Hz.

A diagram of the electrical circuit of a typical spinner magnetometer is shown in figure 8.8. The phase of the induced voltage is found by comparison with a reference voltage derived from the spinner. In the figure, a permanent magnet is fixed to the axis of the spinner and induces a voltage in a second pick-up coil. The voltages from the two coils are amplified and are then compared in a detector circuit, the output of which depends on the difference of phase between the two signals. If the reference voltage is of constant amplitude, the maximum output will be obtained when the two signals are in phase and will depend only on the rock signal provided the reference signal is of much greater amplitude. The direction of the rock sample for which the signals are in phase may be found by rotating the reference coil until the detector output is zero (the most sensitive setting), when the rock signal will be 90° out of phase with the reference signal and the angle between the rock coil and the reference coil will be equal to 90° plus the angle between the direction of the permanent magnet and the direction of the moment of the rock specimen.

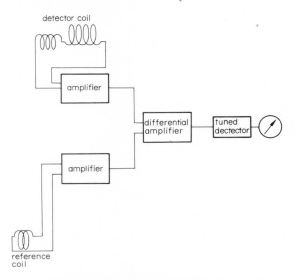

Figure 8.8 Electrical circuit of spinner magnetometer

Spinner magnetometers can measure fields of the order of 10^{-13} T corresponding to a magnetization of 10^{-10} T in a 1 cm^3 cube at a distance of 10 cm.

Two methods are available for demagnetizing a rock and separating the magnetization into hard and soft components. In the one, the rock is placed in an alternating magnetic field of which the amplitude is readily reduced to zero. From the way in which the residual magnetization depends on the initial alternating field, it is possible to separate the primary and secondary components (see figure 8.9). Since the directions of the components of magnetization are in general arbitrary, the rock must be placed in three mutually perpendicular directions in the field, and it is usual to rotate the sample simultaneously about two perpendicular axes at a rate that is fast compared with the rate of reduction of the alternating field. An electrolytic resistance in the alternating current circuit enables the current to be reduced smoothly and slowly. The demagnetization must be carried out in a region in which there is no steady

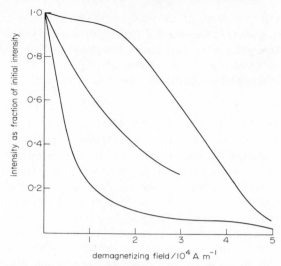

Figure 8.9 Demagnetization curves of rock samples

field, for a steady field in the presence of the alternating field will induce a remanent magnetization in the rock.

The other method of demagnetization is to heat the specimen and to cool it in the absence of a steady field. The specimen is placed in an electric oven designed so that the heating current produces no net field and is surrounded by magnetic shields and Helmholtz coils so that the steady field may be reduced to zero. The specimen must be placed in an inert atmosphere because, if it is oxidized, misleading results may be obtained. It is best to design the furnace so that it may be placed below a magnetometer, enabling the magnetization to be measured

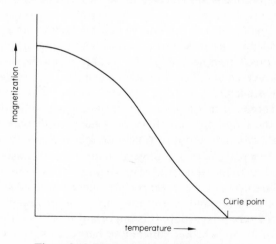

Figure 8.10 Thermal demagnetization curve

without cooling the specimen. In one design the specimen can be rotated about perpendicular axes while in the furnace. Specimens should not be large lest they take too long to reach thermal equilibrium, but one with a volume of 10 cm³ will reach a steady temperature in some 3 min. An example of the behaviour of a specimen under thermal demagnetization is shown in figure 8.10.

8.5 The collection and statistics of rock samples

When samples of rocks are taken for magnetic measurements, the first requirement is that the orientation relative to the present geographical coordinates—north and vertical—should be accurately recorded, for, when that is done, it is possible to relate the orientation to coordinates in the rock and so to the attitude of the rock at the time at which it was formed. At first, specimens were taken from surface exposures of the rock and a line was marked on the specimen, the orientation of which was then measured. Surface specimens are, however, often to be avoided because the action of rain and the atmosphere (weathering) alters the minerals in the rock and destroys the original magnetization to a greater or less extent. In certain climates the surface rocks are liable to magnetization by lightning, and so on both counts it is preferable if possible to take samples from a few metres below the surface; drills have been developed to take specimens from depth while preserving the orientation. Rocks show considerable variation in their magnetic properties so that samples should be taken from as many locations as possible in the same formation of rock, with a number of samples at each location. Samples should also be taken from rocks which maintain the attitude in which they were formed—that is to say, the beds of sedimentary rocks should lie horizontally, as should lava flows, and other igneous rocks such as dikes should be intruded into undisturbed sedimentary rocks.

When the rock specimens are received in the laboratory, small pieces are cut from them of a size suitable for measurements with a magnetometer. The equipment used for cutting up the rock should be non-magnetic so that, if pieces wear off and remain imbedded in the sample, they do not produce a spurious magnetization. Another precaution necessary with weakly magnetized samples is that they should be preserved in a field-free region lest they be remagnetized in the Earth's field in the store.

The basic measurement in studies of the remanent magnetization of rocks is the direction of the magnetization. It is convenient to display the results by plotting the points in which the directions intersect a unit sphere, and to project the points on to a plane map by a projection in which a hemisphere of the unit sphere is projected on to the equatorial plane (figure 8.11). Commonly, as already mentioned, measurements of the direction of magnetization of different samples of a rock do not coincide and, correspondingly, the projected points do not coincide but lie in a scattered group. It is important to analyse the results statistically in order to know whether differences between the directions of magnetization of rocks of different ages or of rocks of the same age but from different places are merely the consequence of the scatter in the measurements or whether they show real differences from place to place or from age to age. The treatment usually followed is due to R. A. Fisher (1953). If a single quantity, say a length or a voltage is measured, it is usually found that in consequence of errors of measurement the chance, $P(x)\,\mathrm{d}x$, of obtaining a value x is given by

$$P(x)\,\mathrm{d}x = \frac{1}{(2\pi)^{1/2}\sigma}\,\mathrm{e}^{-(x-x_0)^2/2\sigma^2}\,\mathrm{d}x$$

where x_0 and σ are parameters that describe the distribution of the observations.

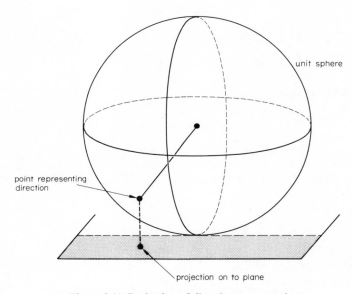

unit sphere

point representing direction

projection on to plane

Figure 8.11 Projection of directions on to a plane

The best estimate of x_0 is the mean $\bar{x}$ of N measured values x, while the best estimate of σ is s, the standard deviation of the observations from the mean:

$$s = \sum_i (x_i - \bar{x})^2/(N-1)$$

Fisher supposed that the corresponding law of chance for directions is that the probability, $P(\psi)\,\mathrm{d}\psi$, of obtaining a result that departs by an angle ψ from the true direction is

$$P(\psi)\,\mathrm{d}\psi = \frac{\kappa}{2\sinh\kappa}\,\mathrm{e}^{\kappa\cos\psi}\sin\psi\,\mathrm{d}\psi$$

where $1/\kappa$ is a measure of the scatter of observations. If $\kappa = 0$, directions are scattered uniformly over the unit sphere, while, if κ is large, directions lie in a tight cluster.

The best estimate of the mean direction is obtained by making the sum of the values of $\cos\psi$ a maximum. Let (λ, μ, v) be the (unknown) direction cosines of the true direction and (l, m, n) those of an individual measured direction. Then

$$\sum \cos \psi = \sum (l\lambda + m\mu + nv) = \lambda \sum l + \mu \sum m + v \sum n$$

and the maximum occurs when

$$\frac{\lambda}{\sum l} = \frac{\mu}{\sum m} = \frac{v}{\sum n}$$

The length of the resultant vector in the direction of the mean is given by

$$R^2 = (\sum l)^2 + (\sum m)^2 + (\sum n)^2$$

and its direction cosines by

$$\lambda = \frac{\sum l}{R}, \quad \mu = \frac{\sum m}{R}, \quad v = \frac{\sum n}{R}$$

Significance tests are constructed on the basis of Fisher's results that the probability, $P(c)$, of obtaining a value of $\cos\psi$ less than c in a sample of N directions is

$$P(c) = \left(\frac{N - R}{N - Rc}\right)^{N-1}$$

If N is large, the estimate of κ is

$$\hat{\kappa} = \frac{N - 1}{N - R}$$

It is then possible to set up tests to decide whether the means of two sets of observations are really distinct or whether the difference between them could be due to the scatter of observations within each group. The careful application of such tests is of great importance in the analysis of the results of measurements of the magnetization of rocks.

One of the causes of scatter in the observed directions of rocks of the same formation is the secular variation, whereby the geomagnetic pole moves in a roughly circular path about the geographic pole with a radius of some 15°. At the present time the path appears to be completed in about 10^3 y, and so, if a formation of rocks was formed in a longer time, the directions of magnetization for rocks of that formation would correspond to poles lying anywhere along the secular variation path.

8.6 Reversed magnetization

The magnetization of many recent lava flows, of ages within perhaps 5 My, are found to be parallel or nearly parallel to the present direction of the magnetic field of the Earth at the site but, while some have the direction corresponding to induction in the present field, they are often interspersed with others that have the opposite direction, as though they were induced by a field opposed to the present field. Reversed magnetization of lavas has been known for over 50 years, and more recently, since it has been possible to take undisturbed cores of sediments from the deep oceans and to measure the weak detrital remanent magnetization that they may possess, it has been found that oceanic sediments also show

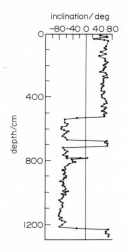

Figure 8.12 Reversals of magnetization of oceanic sediments

alterations of normal and reversed directions, a normal direction being that in which, allowing for secular variation, a rock would be magnetized by the present field of the Earth. An example is shown in figure 8.12.

Because assemblages of minerals with suitable properties can be magnetized in the opposite direction to the magnetizing field and, because some examples at least are known of rocks that behave in that way, it is not obvious that the occurrence of rocks with reversed magnetization necessarily means that the magnetic field of the Earth has at times in the past been of the opposite direction to the present field. In fact, most rocks that show reversals of magnetization are not found to have the special assemblages of minerals that would lead to magnetization in the opposite direction to the magnetizing field, but there are two much stronger pieces of evidence that suggest that the Earth's field has reversed, although some results are difficult to explain by this hypothesis alone.

When lavas are injected into rocks near the surface of the Earth they heat up the surrounding rock, sometimes to a temperature above the Curie point of minerals in the rock. Thus both the lava and the rocks into which it is injected will become magnetized as they cool through their respective Curie points. The minerals in lavas and, for example, sandstones, are different and, if a reversed direction in the lava were due to self-reversal, it would be very surprising if that in the baked rocks in contact with the lava were also due to self-reversal. If the baked contact rocks are found to be magnetized in the same reversed direction as the lava, there is a strong presumption that the Earth's field was also reversed. J. H. Hospers first applied this test in his work in Northern Ireland and Iceland (1951) and it has since been confirmed elsewhere.

The other evidence that points very strongly to a reversal of the Earth's field is that reversed magnetization occurs in rocks of the same ages throughout the world. It was seen in chapter 6 that the potassium–argon method of finding the ages of rocks has been developed to such a sensitivity that it can now be applied to lavas that contain little potassium and that are only about 1 My old so that the argon content is also very low. The upshot of observations on young lavas is that all lavas known to have been formed in the present field of the Earth have the normal direction of magnetization and that, over the past 4 My, all rocks of the same age are either normal or all reversed. The data from deep sea cores support the conclusion that the Earth's field has reversed, for they exhibit detrital remanent magnetization, a mechanism quite different from the thermal remanence of lavas and, in addition, and most important, they show the same world-wide sequence of normal and reversed phases. Lava flows do not usually cover a long interval of time, but complete sequences over the last 5 My can be obtained in a single deep sea core; radioactive dating cannot be carried out on the cores but the distance of a band of magnetization from the surface gives a relative scale of time (provided allowance is made for compaction) that can be correlated with the absolute but partial sequences in lava flows.

It is now possible to draw up a table of periods in which the Earth's field was reversed. It is convenient to give names to the principal periods and they, together with the ages, are shown in table 8.2. The periods are by no means of equal length and, within a period of predominant polarity, a number of short episodes of the other polarity may often occur.

Table 8.2   Reversals of the geomagnetic field

Epoch	Event within epoch	Start of epoch (My)	Date of event (My)
Brunhes (N)		0·75	
	Jaramillo (N)		1
	Olduvai		2
Matayama (R)		2·5	
	Mammoth (R)		3·1
Gauss (N)		3·3	
Gilbert (R)		4·0	

There is one group of observations which is not explained by the hypothesis of the reversals of the field and of which the significance is not yet understood. There are many examples of rocks with reversed magnetization which, although they are of the same type of rocks as adjacent ones with normal magnetization, yet show some small differences. The usual difference where one is found is that in the reversed rock the iron-bearing minerals are more highly oxidized than in the associated normally magnetized rocks; in addition, there may be more titanium in the rocks with the reversed magnetization and there may be differences in the sizes of the grains of some minerals. These results are not yet understood.

If the Earth's field reverses, it may be asked, how it does so? Does the field maintain the same intensity but rotate through 180° or does it decay to zero and then grow again in the opposite direction. To answer this question by observation is very difficult because the time within which the field changes sign seems to be very short (about 10 000 y) and the chance of suitable lavas being formed at just the right time is small. Nonetheless, some instances have been discovered and the indications so far are that the main dipole field of the Earth decays to zero and then grows again in the opposite direction, so that the intensity of the field while a reversal is taking place is much less than before or after.

8.7 The ancient field and the movements of continents

The results of measurements on the magnetization of rocks considered in this section are those that critical tests, particularly those involving progressive demagnetization of samples, have shown to correspond to the magnetization originally induced in the rock at the time of formation. It is of course impossible to be certain of any interpretation that involves extending inferences from laboratory observations of magnetization back over perhaps hundreds of millions of years, but the general consistency of the results of interpreting the hardest magnetization of a rock in terms of the Earth's field at the time of the formation of the rock does indicate that some reliance may be placed upon them. At the same time, the results must be treated with caution, not only because of the long interval of time over which an extrapolation has to be made but also because of the scatter often found in the observations, scatter that sometimes results in apparently significant differences between observations which nonetheless are due to the scatter. Careful statistical examination of the results of measurements of the magnetization of rocks is therefore most important.

When one examines the magnetization of rocks from say 200 My ago, one may well find that the directions are so different from the direction of the present field that even a normal or reversed direction cannot be identified. It is clear that either the direction of the pole of the Earth's field has changed or the site of the rocks has moved relative to the rest of the Earth. It is now thought (see chapter 9) that the mean direction of the pole of the Earth's field is parallel to the spin axis of the Earth, which is at any time very close to the axis of greatest moment of inertia. Thus, when we speak of a movement of the pole of the magnetic field of the Earth, we are thinking of a movement of the crust as a whole relative to the axis of greatest moment of inertia, that is relative to the equatorial bulge of the Earth, for the axis of greatest inertia is perpendicular to the equatorial bulge.

The magnetization of a rock is commonly interpreted in terms of the position of the pole of the dipole field in which it would have been produced. The plane

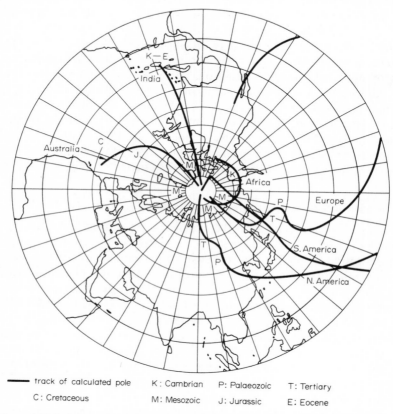

Figure 8.13 Polar wander curve (for the ages of the geological epochs, see table 6.3)

in which the pole lies must contain the direction of magnetization and the radius vector of the site from the centre of the Earth, while the angular distance of the site from the direction of the pole is calculated from the inclination of the magnetization of the rock using equation (8.5). Positions of poles found in this way are commonly displayed by plotting them on a stereographic projection.

If the poles found from the magnetization of recent rocks were plotted upon a stereographic projection, they would all cluster around the north geographic pole for normally magnetized rocks, and around the south pole for rocks with reversed magnetization, and the scatter of the poles about the two directions would correspond to the wander of the magnetic poles in the course of secular variation. The poles found from rocks of the same age within a single continent cluster in the same way about directions 180° apart although the directions depart

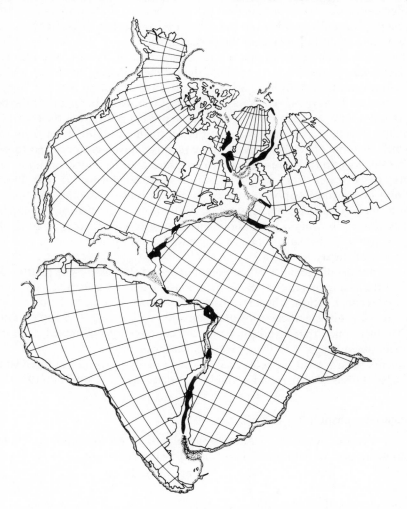

Figure 8.14 Continental reconstruction (after Bullard *et al.*, 1965)

more and more from the present geographical poles as the rocks get older. As figure 8.13 shows, the poles found from rocks in different continents do not in general coincide and, for the older rocks, the differences appear to be statistically significant despite the spread of the observations. In fact, as the age increases, so in general do the angular distances between the poles found from the different continents, up to an age of 300 My. Prior to that period, the poles obtained from different continents appear to move around the Earth as a group. The results can be expressed graphically as paths of polar wandering, which show the tracks of the poles found from the different continents as time passes. Such polar wander paths are shown in figure 8.13. It will be seen that up to some 300 My years ago (Carboniferous Period) the paths are parallel but thereafter they come closer together until they meet at the present time.

The results summarized in figure 8.13 suggest that the crust of the Earth as a whole has moved relative to the pole of the magnetic field of the Earth and also that continents have moved relative to each other. Prior to 300 My ago, the continents could have moved as a rigid block but, because the paths of the poles do not coincide when they are found from the present position of the continents, it follows that the relative positions of the continents were different from those in which they now find themselves. Subsequently, movement occurred to bring the continents into their present relative positions and the course of those movements can be traced from the paths of polar wander during the past 300 My. Not only do continents appear to have moved relative to each other but they also seem to have rotated in some instances. A rotation is revealed by the apparent pole position following a path about the continent as a centre; an example is Spain which seems to have rotated relative to the rest of Europe from a position in the Bay of Biscay.

The positions of the poles of the past magnetic fields enable the relative positions of the continents in the past to be reconstructed and a map of a commonly accepted reconstruction, which is, however, based on different evidence, is shown in figure 8.14. The significance of this map will be discussed in chapter 10 but it may be mentioned that it indicates that the continents have moved from more compact assemblages into their present locations—the idea of the drift of the continents put forward on geological evidence many years ago by the German geologist, Alfred Wegener, and strongly supported by A. L. du Toit (1937).

Examples for chapter 8

8.1 The directions of magnetization of samples of a rock formation at 10 neighbouring sites were found to have the following directions.

Lat.	Long.	Lat.	Long.
57°N	82°E	58°N	81°E
55	83	57	83
56	86	54	76
57	81	56	85
59	78	52	80

Calculate the direction cosines of each direction and estimate the mean direction and the coefficient of dispersion, K. How many directions, according to Fisher's theory, do you expect to lie within 3° of the mean direction and how many actually so lie?

Note. The direction cosines of a direction given by the colatitude θ and the longitude ϕ are

$$l = \sin \theta \cos \phi$$
$$m = \sin \theta \sin \phi$$
$$n = \cos \theta$$

8.2 The mean direction of magnetization of a certain rock formation at a site in lat. 30°S long. 20°E has an inclination of 30° and bears 50°W of N. Calculate the position of the pole of the Earth's field (assumed to be a dipole) at the time of magnetization.

Note. The angular distance between the site and the pole follows from equation (8.5). Let the present geographic N pole be N, the ancient pole P and the site S. Let the geographical colatitude and longitude of P be θ_p, ϕ_p. Then, in the spherical triangle NSP, the sides NS and SP and the angle NŜP are known and so NP may be found from the cosine rule and hence the angle SN̂P from the sine rule.

Further reading for chapter 8

Bullard, E. C., Everett, J. E., and Smith, E. G., 1965, 'The fit of continents around the Atlantic', *Phil. Trans. Roy. Soc., Ser. A*, **258**, 41–51.

Fisher, R. A., 1953, 'Dispersion on a sphere', *Proc. Roy. Sci., Ser. A*, **217**, 295–305.

Irving, E., 1964, *Palaeomagnetism* (New York, London, Sydney: Wiley), ix + 399 pp.

du Toit, A. L., 1937, *Our Wandering Continents* (Edinburgh: Oliver and Boyd), 396 pp.

CHAPTER NINE

The main magnetic field of the Earth

'A marvel and a mystery to the world.'
Longfellow, *Michelangelo*

9.1 Introduction

Although the magnetic field of the Earth is predominantly that of a dipole at the centre, there are many additional components that are of interest in themselves and must also be taken into account in a study of the main field. The field at the surface undergoes daily and other short-period variations which are caused by the electrical currents circulating in the ionized gases around the Earth and which can be divided into two sorts—those with a steady daily cycle, or quiet-day variation, denoted by Sq, and the larger, more erratic ones called 'storm-time variations' (Dst) which sometimes begin suddenly with a 'sudden commencement' (ssc) and die away slowly over some 48 h. The quiet daily variations arise from the rotation of the Earth and the main magnetic field in a sheath of ionized gas which maintains a constant orientation with respect to the Sun; while the storm-time variations are caused by the arrival of charged particles in the neighbourhood of the Earth following an outburst on the Sun. The amplitude of the Sq variation may reach about 4×10^{-8} T and the peak of a Dst pulse may reach 5×10^{-8} T. This book is about the physics of the solid Earth but daily variations affect the measurement of the main field and the short-period variations are used to determine the electrical conductivity of the Earth. The main field also undergoes a slow variation which to a first approximation may be represented as a rotation of the dipole at the centre of the Earth; the study of the secular variation has illuminated the study of the origin of the main dipole field and is related to other geophysical phenomena as well, so that the description of the secular variation is a principal aim of the study of the magnetic field of the Earth.

The methods for the measurement of the magnetic field of the Earth for long depended upon the mechanical forces acting on permanent magnets, following principles established by Gauss, but in recent years, electronic methods have become more important, especially those depending on the behaviour of atoms in a magnetic field. Not only are modern measurements more accurate than the traditional ones, but they can be made automatically and therefore more frequently and under remote control, in particular in artificial satellites and space probes, and indeed it is now possible to measure the magnetic fields of planets from space vehicles.

A detailed description of the geomagnetic field is still essential in navigation, and to interpret other observations in geophysics, especially the behaviour of charged particles around the Earth and of cosmic rays, but the emphasis here is placed upon the physical processes by which the field is generated and how they may be related to the structure of the Earth as inferred from seismic and other studies. The Earth is unique among the neighbouring terrestrial planets in possessing a general magnetic field and the relation of this fact to the physics of the other planets will also be considered.

9.2 The measurement of magnetic field

Until the last ten or twenty years, all measurements of the magnetic field of the Earth depended on the classical principles of electromagnetism, many of which were established by Gauss. Thus the dip and inclination were found from observations of delicately suspended permanent magnets, while variations in the intensity were also often found from observations of the change in angle of a magnet suspended at right angles to the field by a torsion fibre the torque exerted

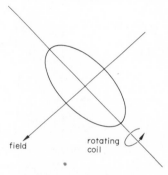

field   rotating
             coil

Figure 9.1 Earth inductor

by which balanced the couple exerted on the magnet by the field of the Earth. The total field or any component can be determined from the voltage induced in a coil rotated in the field, the component that is measured being that perpendicular to the axis of rotation (figure 9.1). The inductor may also be used to measure the dip by finding the orientation for which the voltage is zero, and is indeed to be preferred to the dip needle. The most refined examples of classical instruments are the quartz horizontal magnetometer (QHM) and the vertical magnetic balance (BMZ). The former comprises a permanent magnet of high coercive force suspended, with its axis horizontal, from a vertical quartz torsion fibre anchored at both ends. One end of the fibre is rotated until the magnet is at right angles to the horizontal component of the Earth's field, and the change in rotation needed to keep the magnet in a constant alignment is a measure of any change in H. The BMZ consists of a magnet integral with a knife edge support and counter balance weight adjusted to balance with its axis horizontal. If Z, the vertical component of the field, changes, the magnet is restored to the horizontal by altering the position of an auxiliary magnet placed below it so keeping constant the total field of the Earth and the auxiliary magnet at the balanced magnet. The QHM and BMZ are not absolute instruments but, through careful design and construction and the use of very hard magnetic materials, they behave in a highly stable manner.

The QHM will measure changes of the horizontal field to 10^{-9} T and the BMZ will measure changes of the vertical field to 10^{-9} T as well.

Nowadays most observations of the Earth's magnetic field are made with instruments that use the magnetic properties of atomic nuclei or with electronic devices that depend on the saturation of the magnetization of ferromagnetic materials. The proton precession magnetometer makes use of the magnetic moment of the proton, which classically may be considered to have angular momentum, M, and a magnetic moment, μ. Consider a proton with its magnetic moment μ at an angle θ to the applied field B (figure 9.2). The couple exerted on the proton is $B \wedge \mu$ or $B\mu \sin\theta$ in the direction perpendicular to B and μ. Since a couple is equal to a rate of change of angular momentum, the change of angular

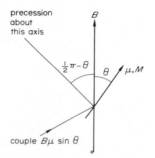

Figure 9.2 Proton precessing in a magnetic field

momentum of the proton in a time δt is $B\mu \sin\theta \, \delta t$. Then to conserve the total angular momentum the proton rotates by an angle $\delta\chi$ given by

$$M\delta\chi = B\mu \sin\theta \, \delta t$$

about the axis perpendicular to M and in the plane of B and M.

The rate of rotation about the direction of B is accordingly

$$\frac{1}{\sin\theta} \frac{\mathrm{d}\chi}{\mathrm{d}t} = \frac{B\mu}{M}$$

The ratio μ/M is known as the *gyromagnetic ratio* of the proton. In a sample of water under normal conditions, the protons all tend to line up with the applied field but, at the same time, thermal motions cause them to move in a random way so that there is no net magnetization. If, however, a strong field is applied to the sample of water, the protons will be aligned with this field and, then if the polarizing field is suddenly removed, the protons will precess all together around the Earth's field for some twenty seconds before thermal movements destroy the coherence of the motion of the different protons. The rotating protons will induce a voltage in a coil during the time they are all moving together and the frequency of the voltage is a measure of the Earth's field according to the formula given above. The magnitude of the voltage depends on the angle between the axis of the pick-up coil and the Earth's field but the frequency depends on the field alone. The behaviour of the proton precession can equally, and perhaps more simply, be described in quantum-mechanical terms, for the proton can be in one of two states in a magnetic field; in one the magnetic moment is parallel to the field and in the other it is anti-parallel. The difference of energy between the two states is $2B\mu$. According to the principles of quantum mechanics, transitions can take place from the state of higher to the state of lower energy with the emission of radiation at the frequency v given by

$$hv = 2B\mu$$

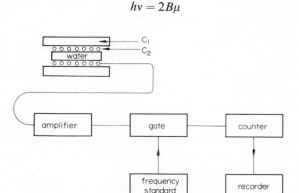

Figure 9.3 Proton precession magnetometer

Since the angular momentum of the proton, which has a spin of $\frac{1}{2}$, is $\frac{1}{2}h$ this formula is the same as that given previously.

A diagram of the circuits of a proton precession magnetometer is shown in figure 9.3. C_1 is the polarizing coil through which a short strong pulse of current may be passed to produce a field to align the protons, and C_2 is the detector coil in which the precession voltage is induced. After amplification, the induced signal is fed to an electronic counter with which the frequency is measured.

The gyromagnetic ratio of the proton has been measured in a number of laboratories such as the National Physical Laboratory in England and the National Bureau of Standards in the U.S.A. because the proton magnetometer may be used in reverse to establish a standard magnetic field, and hence a standard current, from measurement of the frequency of precession in that field. The best result at present is $4 \cdot 258 \times 10^7$ Hz T^{-1}.

The frequency of precession in a field of some 5×10^{-5} T, the Earth's field, is about 2 kHz.

Various circuits have been constructed that manipulate the signals from the detector coil so that a direct reading of the field strength is obtained.

In practical instruments, a current of about 100 A is passed through the polarizing coil and, after it has been switched off (in a time less than 50 μs), the electronic circuits start to measure the frequency of the precession signal. The time for which that signal lasts depends upon the time for which the motion of the protons remains coherent, which in turn depends on the coupling between the protons and other nuclei in the liquid; in practice water and benzene give long enough decay times. Other effects reduce the length of the signal, in particular, any lack of uniformity in the field but, in practice, times of some 20 s are common. In that time the frequency can be measured to the equivalent of better than 10^{-9} T (1γ). The direction of the detector coil makes an arbitrary angle with the direction of the Earth's field and so the signal could at times be very weak; practical instruments for survey use are therefore usually provided with two coils at right angles, or else with coils wound in a special configuration to minimize the variation of the signal with orientation.

Proton precession magnetometers have proved very suitable for measuring the total field when towed behind a ship or when trailed from an aircraft or when carried in a rocket or satellite; the fact that the measured intensity does not depend on orientation is of particular value when the instrument is on a moving platform.

The main disadvantage of the proton precession magnetometer is that the observations are made in a cycle of some 20 to 30 s; it is therefore unable to detect variations of the field of shorter period and is somewhat inconvenient for use in a space vehicle. Optical pumping magnetometers overcome the difficulty and record the total field continuously.

The rubidium vapour magnetometer is the optical pumping magnetometer most commonly used. In it, resonance radiation (at a wavelength of 800 nm) that originates from transitions between the ground state and first excited state of rubidium is generated in an electrical discharge lamp and then passes through a circular polarizer, after which the circularly polarized light falls on a cell containing rubidium vapour. In the normal way, the resonance radiation would be strongly absorbed by rubidium atoms in the ground state. However, the vapour is in the Earth's magnetic field and is also subject to an oscillating magnetic field and, if the latter has the correct frequency, the numbers of atoms in the states between which the optical transitions take place are changed by induced transitions between neighbouring Zeeman levels; then the absorption of the light is altered and may be used to maintain the oscillating field at the correct frequency which is then a measure of the steady Earth's field.

The relevant part of the energy level diagram of rubidium is shown in figure 9.4. The quantum number J for the total electronic angular momentum in both the ground state and excited state is $\frac{1}{2}$ and, since the spin of the nucleus of rubidium-85 is $\frac{5}{2}$, the total quantum number F, equal to $I \pm \frac{1}{2}$, can take the values 3 or 2. Consider the transition between the levels $F = 2$ in the excited state and $F = 3$ in the ground state. When a magnetic field is applied, the level $F = 2$ splits into 5 magnetic sub-levels with $m = -2, -1, 0, +1$ and $+2$ and the $F = 3$ level splits into 7 with $m = -3, \cdots, 0, \cdots, +3$. The difference of frequency between the latter are all given to first order by

$$v = 4\cdot67 \times 10^9 \text{ Hz T}^{-1}$$

Thus v is about 2330 MHz in the Earth's field.

As with the proton precession magnetometer, the frequency of oscillation is independent of the orientation of the magnetometer to the Earth's field but the intensity of the oscillating signal does depend on the orientation, in fact it is proportional to $\sin 2\theta$, where θ is the angle between the direction of the field and the direction of the beam of light. Thus the magnetometer ceases to work when θ is 0° or 90°, and in artificial satellites four gas cells are provided to reduce the resulting insensitive zone to a minimum.

The rubidium magnetometer has the advantage over the proton magnetometer

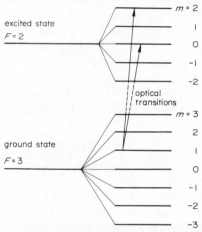

Figure 9.4 Part of energy-level diagram of rubidium-85

that it runs continuously and that its speed of response to a change in magnetic field is high so that it may be used to follow rapid changes of the field. The characteristics of the two magnetometers are such that the proton instrument is used for absolute laboratory measurements and when towed behind ships at sea, whilst the rubidium magnetometer is used in satellites and space probes to measure the total field, and in observatories to follow rapid changes of the field.

Both magnetometers may be provided with auxiliary coils to enable the orthogonal components of the field to be determined. Suppose that the three components are denoted by H_x, H_y and H_z, and suppose that H_x is to be found. Let an auxiliary field be applied in the x direction, by means of a current through a pair of Helmholtz coils, first in the same direction as H_x and then in the opposite direction. If the magnitude of the auxiliary field is h_x, the measured total fields are given by

$$H_1^2 = (H_x + h_x)^2 + H_y^2 + H_z^2$$

and

$$H_2^2 = (H_x - h_x)^2 + H_y^2 + H_z^2$$

Thus

$$H_1^2 - H_2^2 = 2H_x h_x$$

from which, knowing h_x, H_x may be found.

The proton magnetometer will measure the Earth's field to about 10^{-9} T; the rubidium magnetometer is somewhat more sensitive.

A third magnetometer widely used, particularly for surveying, is the fluxgate magnetometer which depends on the properties of ferromagnetic materials. A core of high-permeability alloy is wound with two coils through which equal alternating currents are passed in opposite directions, the currents being large enough that the magnetic fields that they produce saturate the cores for most of a cycle. The whole core is wound with a single coil in which a voltage is induced by the changing flux that links it (figure 9.5). If now the core is subject also to a steady field, that of the Earth in particular, then because the relation between B

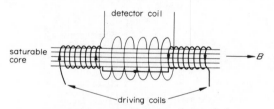

Figure 9.5 The fluxgate magnetometer

and H is non-linear in the core the variation of flux generated by the one half of the coil is not the mirror image of that generated by the other half and there is a net positive or negative flux (according to the direction of the steady field) through the detector coil. The rate of change of this net flux is equal to the voltage induced in the detector coil and is characterized by the fact that, as indicated in figure 9.6, it contains the second harmonic of the alternating inducing current. The magnitude of the second harmonic component is proportional to the steady field and the phase gives the direction of the steady field. The signal in the detector coil may be amplified by a circuit that responds only to the second harmonic (phase-sensitive detector).

Because the fluxgate magnetometer responds to the component of the field parallel to the core, it may be used to measure any component of the field. There are two ways of making use of this property. In magnetic surveys made from aircraft for geophysical prospecting and for other geological purposes, the aim is to measure the total field and so the main magnetometer is associated with two others that turn it, by means of a servo-mechanism, to be parallel to the resultant field so that it measures the total field. Similar instruments have been used in satellites. In the other form of the instrument, developed in Canada for use in land surveys, three cores at right angles measure the three orthogonal components of the field, being set up by the surveyor, with the aid of a theodolite mounting, so that the components measured are the vertical and the horizontal ones in and at right angles to the meridian.

By making an appropriate selection from the instruments that have been described, the total field of the Earth, the Moon or a planet may be measured in absolute terms to better than 10^{-9} T, a particular component on the Earth may be measured to about 10^{-9} T and variations in the field may be measured to about 10^{-10} T at frequencies as high as 1 kHz.

Geological structures can be investigated, much as they are in gravity surveys, by study of the variations in the local magnetic field caused by related changes in magnetization. In principle, the problem is similar to gravity survey but is more complicated. In the first place, either the total force or the vertical component

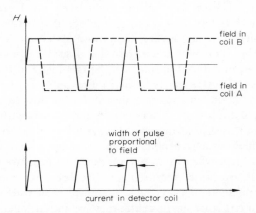

Figure 9.6 Changes of flux in the fluxgate magnetometer

may be measured, but each must be measured in the presence of the daily variation, which must therefore be recorded at a fixed station throughout a survey. The Earth's field varies with latitude just as does the gravity field and so a correction must be made much as in gravity survey. For an axial dipole,

$$Z = \frac{2S\cos\theta}{r^3}$$

and

$$H = \frac{S\sin\theta}{r^3}$$

The changes with colatitude are

$$\frac{\partial Z}{\partial \theta} = -\frac{2S\sin\theta}{r^3} \quad \text{and} \quad \frac{\partial H}{\partial \theta} = \frac{S\cos\theta}{r^3}$$

or

$$\frac{\partial Z}{\partial \theta} = -2H \quad \text{and} \quad \frac{\partial H}{\partial \theta} = \tfrac{1}{2}Z$$

the units in each case being tesla per radian. In terms of tesla per degree,

$$\frac{\partial Z}{\partial \theta} = -\frac{\pi H}{90}, \quad \frac{\partial H}{\partial \theta} = \frac{\pi Z}{360}$$

In latitude $45°$ where Z is about 4×10^{-5} T and H about 2×10^{-5} T, $\partial Z/\partial \theta$ is about $1 \cdot 3 \times 10^{-6}$ T deg^{-1} or $1 \cdot 2 \times 10^{-8}$ T km^{-1} and $\partial H/\partial \theta$ is about $0 \cdot 3 \times 10^{-6}$ T deg^{-1} or 3×10^{-9} T km^{-1}, values which should be compared with anomalies of less than 10^{-8} T.

Since the geomagnetic field is that of a dipole whereas the gravity field is that of a point mass, the variation with latitude is relatively much greater but, on the other hand, so may be the anomalies due to geological structure, for the relative differences of magnetization are greater than those of density. The interpretation of magnetic anomalies is more complex than that of gravity anomalies because of effects of magnetization induced in rocks by the main field of the Earth.

Magnetic surveys are extensively used in prospecting: they are especially suitable for locating ferromagnetic ore bodies and other mineralized regions; and it is relatively simple and cheap to survey very large areas with magnetometers towed behind aircraft or ships. Magnetic surveys do not have to be corrected for the Eötvös effect which is such a problem in shipborne gravity measurements (and even more of a problem when attempts are made to measure gravity from an aircraft) and the navigational requirements are less severe than for gravity surveys. Airborne magnetic surveys have therefore been widely applied to the study of the general structure of the crust; they are in particular sensitive to the varying depth of the more basic components of the crust which contain a higher proportion of ferrimagnetic minerals and so acquire a higher magnetization in the field of the Earth than the more acidic components. Magnetic anomalies arise from only the outer layers of the Earth for, at quite moderate depths, the temperature rises above the Curie points of the common ferrimagnetic minerals in rocks and at greater depths the rocks will not be magnetized.

9.3 Surveys of the main field

From the seventeenth century onwards, the main field of the Earth has been studied to obtain an accurate description of the field as it is, at first exclusively for use in navigation, and to attempt to develop a theory of the field so described. Edmond Halley was the first to draw a magnetic chart based on a cruise of 2 years between 1698 and 1700 and he also speculated on the origin of the field and in particular of the secular variation, of which he was aware. The main aim of world-wide surveys was for long to draw up charts of quantities such as declination and inclination and of the rates at which they were changing, and the problem is complicated by the daily and long-term changes undergone by the field which must be measured continuously and allowed for in finding the steady field from measurements at different times in different places. In the last decade, the use of satellites has enabled world magnetic surveys to be carried out at very much greater speed and in much greater detail than previously.

The description of the field is given in terms of the *magnetic elements*, that is to say, the components of the field, the dip or inclination, and the declination. The vertical component of the field is usually denoted by Z and the horizontal component by H. The component in the meridian is denoted by X and that in the east–west direction perpendicular to the meridian by Y. If F is the total field,

$$F^2 = Z^2 + H^2$$

$$= Z^2 + X^2 + Y^2 \qquad (9.1)$$

The inclination, the angle at which the field direction is inclined to the horizontal, is given by

$$\tan I = H/Z \qquad (9.2)$$

so that

$$F = H \cos I \qquad (9.3)$$

while the declination, the angle that the horizontal component makes with the meridian, is given by

$$\tan D = Y/X \qquad (9.4)$$

Until the advent of total field instruments and measurements in aircraft and satellites, the usual practice was to measure the inclination and declination and the horizontal component H, and to derive the other components where necessary. Nowadays it is more usual to measure the total field F from which all other elements may be derived, as described below.

The results of magnetic surveys are commonly presented as charts such as those shown in figures 9.7 to 9.10. Figure 9.7 shows lines along which the dip is the same (isoclinic lines) and figure 9.8 those on which the horizontal intensity is the same (isodynamic lines). Figure 9.9 is a chart of the total field (see example 9.4) and figure 9.10 is a chart of lines along which the rate of change of declination are the same (isoporic lines).

Charts are the most convenient way in which to present the geomagnetic field to the navigator but both in compiling them and in analysing the field it is better to have data in numerical form. Since the time of Gauss it has been the practice to give the components of the field as a series of spherical harmonics (see chapter 2).

The magnetic field may be derived from a scalar potential, V, by the equation

$$\boldsymbol{B} = -\text{grad } V$$

If positions on the surface of the Earth are given in spherical polar coordinates

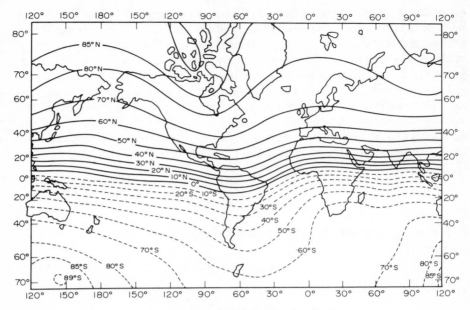

Figure 9.7 Chart of inclination (isoclines)

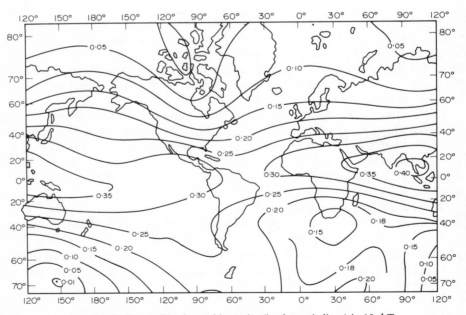

Figure 9.8 Chart of horizontal intensity (isodynamic lines) in 10^{-4} T

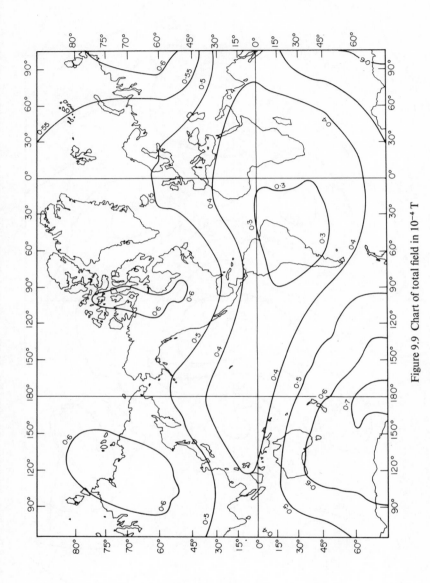

Figure 9.9 Chart of total field in 10^{-4} T

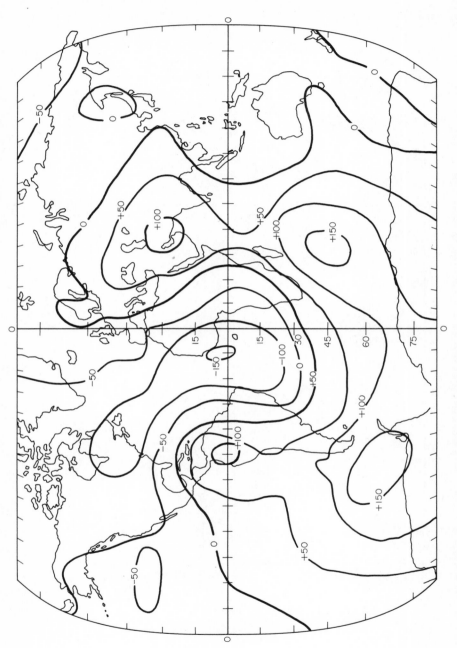

Figure 9.10 Chart of variation of vertical field (isopores). Secular change, 1922–5, vertical intensity. Changes in 10^{-9} T y^{-1}

(r, θ, ϕ) where, as usual, r is the radius vector, θ the colatitude and ϕ the longitude, then the components of the field are

$$\left.\begin{array}{c} Z = -\partial V / \partial r \\[2mm] X = -\dfrac{\partial V}{r\,\partial\theta} \\[3mm] Y = -\dfrac{\partial V}{r \sin\theta\,\partial\phi} \end{array}\right\} \qquad (9.5)$$

As with the gravitational potential, the magnetic potential is expanded in a series of spherical harmonics, but there is an important difference in that the sources of the geomagnetic field are not all inside the Earth. The general solution to Laplace's equation, which is satisfied by the geomagnetic potential at the surface, comprises two series of terms, those with the radial factor r^n and those with the factor r^{-n-1} (appendix 2). Since there are no sources of the gravity field outside the Earth (apart from the tide-raising potential) the potential cannot contain any part that increases with distance from the centre of the Earth and so the only terms are those with the factors r^{-n-1}. The magnetic field, on the other hand, is in part generated by external currents in the ionosphere which, unlike the tide-raising potential, cannot be neglected to first order, for they may contribute up to one per cent of the surface field instead of the parts in ten million of the tidal potential. Terms with the factor r^n must therefore be included in the expression for the potential, although they all vary rapidly with time, whereas the internal terms contain the steady parts and those that vary slowly as well as rapidly.

The general expression for the potential is then written, following Gauss, as

$$V = a \sum_{n=1}^{\infty} \sum_{m=0}^{m=n} \left(\frac{a}{r}\right)^{n+1} (g_{ni}^m \cos m\phi + h_{ni}^m \sin m\phi) P_{ni}^m(\cos\theta)$$

$$+ a \sum_{n=1}^{\infty} \sum_{m=0}^{m=n} \left(\frac{r}{a}\right)^{n} (g_{ne}^m \cos m\phi + h_{ne}^m \sin m\phi) P_{n}^m(\cos\theta) \qquad (9.6)$$

where the coefficients g_{ni}^m and h_{ni}^m refer to the internal sources and g_{ne}^m and h_{ne}^m to the external sources; all may vary with time.

However, in considering the main field, the external sources may be ignored and the potential written as

$$V = a \sum_{n=1}^{\infty} \sum_{m=0}^{m=n} \left(\frac{a}{r}\right)^{n+1} (g_{ni}^m \cos m\phi + h_{ni}^m \sin m\phi) P_{n}^m(\cos\theta) \qquad (9.7)$$

On differentiating in the appropriate direction and putting the radius r equal to the surface radius a, it will be found that

$$Z = -\frac{\partial V}{\partial r}\bigg|_{r=a}$$

$$= -\sum_{n=1}^{\infty} \sum_{m=0}^{m=n} (n+1)\,(g_{ni}^m \cos m\phi + h_{ni}^m \sin m\phi)\,P_n^m(\cos\theta) \qquad (9.8a)$$

$$X = -\frac{\partial V}{r\,\partial\theta}\bigg|_{r=a}$$

$$= \sum_{n=1}^{\infty} \sum_{m=0}^{m=n} (g_{ni}^m \cos m\phi + h_{ni}^m \sin m\phi)\,\frac{\mathrm{d}P_n^m(\cos\theta)}{\mathrm{d}\theta} \qquad (9.8b)$$

$$Y = -\frac{\partial V}{r\sin\theta\,\partial\phi}\bigg|_{r=a}$$

$$= \sum_{n=1}^{\infty} \sum_{m=0}^{m=n} m(-g_{ni}^m \sin m\phi + h_{ni}^m \cos m\phi)\,\frac{P_n^m(\cos\theta)}{\sin\theta} \qquad (9.8c)$$

The differential of an associated Legendre coefficient, $\mathrm{d}P_n^m/\mathrm{d}\theta$, and the quantity $P_n^m/\sin\theta$ may be expressed as associated Legendre coefficients of adjacent order, and so both X and Y can be written in the standard form of a series of surface harmonics. Similarly F, equal to $X^2 + Y^2 + Z^2$ may be reduced to a series of surface harmonics (including now a constant term) (see example 9.1). Thus all the field components may be expressed as a series of spherical harmonics and the coefficients g_{ni}^m and h_{ni}^m may be obtained from any one of them, in particular, they may be derived from the total field that is measured from ships, aircraft or satellites.

Having obtained a mathematical expression for the magnetic potential in spherical harmonics from the observations made around the world, it is a straightforward matter to draw a chart of any one of the conventional elements. As a result of satellite observations in particular, a formula for the potential has obtained general agreement internationally; it contains 40 terms, a few of which are given in table 9.1.

The first term in the potential, proportional to $P_1(\cos\theta)$ or $\cos\theta$, is the potential of a dipole placed at the centre of the Earth and directed along the spin axis of the Earth. The addition of the two terms proportional to $P_1^1(\cos\theta)$ or $\sin\theta$ corresponds to a tilt of the dipole axis, which, however, still passes through the centre of the Earth. Other terms correspond to a displacement of the axis of the dipole from the centre of the Earth and the dipole that best fits the field is displaced by some 300 km.

The components of the dipole moment are $a^3\,g_{1i}$ along the polar axis, $a^3 g_{1i}^1$ in the equatorial plane and through the meridian of Greenwich, and $a^3 h_{1i}^1$ in the equatorial plane and at right angles to the meridian of Greenwich.

Table 9.1 Some coefficients in the harmonic expansion
of the geomagnetic field

n	m	g_{ni}^m	h_{ni}^m
1	0	$-3 \cdot 04 \times 10^{-5}$	
2	0	$-0 \cdot 16$	
3	0	$0 \cdot 13$	
4	0	$0 \cdot 10$	
5	0	$-0 \cdot 02$	
6	0	$0 \cdot 14$	
1	1	$-0 \cdot 21$	$0 \cdot 58$
2	1	$0 \cdot 30$	$-0 \cdot 19$
3	1	$-0 \cdot 20$	$-0 \cdot 05$
4	1	$0 \cdot 08$	$0 \cdot 02$
2	2	$0 \cdot 15$	$0 \cdot 02$
3	2	$0 \cdot 13$	$0 \cdot 02$
4	2	$0 \cdot 06$	$-0 \cdot 03$
2	3	$0 \cdot 08$	—
4	3	$-0 \cdot 03$	—
6	3	$-0 \cdot 02$	—
4	4	$0 \cdot 03$	$-0 \cdot 02$

Units. The coefficients are in tesla.
Normalization. The spherical harmonics that the
coefficients multiply are defined so that the integral of
any harmonic over the unit sphere is $(2n + 1)^{-1/2}$
(Schmidt normalization).
See Jensen and Cain, 1962.

Even allowing the main dipole to be displaced from the centre of the Earth, the
fit to the actual field is not very good, as indicated by the appreciable size of, for
example, g_{2i}. It is instructive to subtract the main dipole field and examine the
remainder, as displayed in figure 9.11. The occurrence of a number of closed
loops in the non-dipole field suggests that it could itself be represented as the sum
of fields of dipoles suitably placed in the Earth and pointing in the directions of

Table 9.2 Representation of the geomagnetic field by radial dipoles

	M/a^3 (T)	Colatitude (deg)	Longitude (deg E)
Central dipole	$-7 \cdot 0 \times 10^{-5}$	$23 \cdot 6$	$208 \cdot 3$
Radial dipoles at $0 \cdot 25a$	$1 \cdot 0$	$13 \cdot 7$	$341 \cdot 9$
	$1 \cdot 1$	$46 \cdot 0$	$179 \cdot 9$
	$-0 \cdot 3$	$54 \cdot 9$	$40 \cdot 1$
	$0 \cdot 8$	$77 \cdot 4$	$241 \cdot 7$
	$0 \cdot 3$	$91 \cdot 3$	$120 \cdot 8$
	$-0 \cdot 9$	$139 \cdot 8$	$319 \cdot 3$
	$-1 \cdot 2$	$141 \cdot 1$	$43 \cdot 0$
	$0 \cdot 4$	$102 \cdot 9$	$180 \cdot 1$

See Alldredge and Hurwitz, 1964.
Note. a is the radius of the Earth. The radial dipoles are placed at a
distance of $0 \cdot 25a$ from the centre. The units of the magnetic moments are
T m^3. The unit of M/a^3, the moment divided by a^3, is thus the tesla.

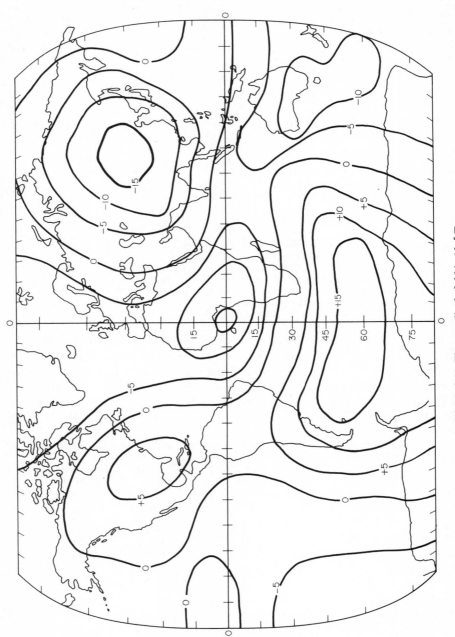

Figure 9.11 The non-dipole field in 10^{-5} T

the respective loops, and a set of such radial dipoles is listed in table 9.2. It will be noticed that the dipoles must be placed quite deep within the core. The representation of the main field in terms of a number of radial dipoles may be physically more significant than the representation in terms of spherical harmonics, for the dipoles would correspond to loops of electrical current roughly tangential to the surface of the core. The coefficients of the spherical harmonics may of course be interpreted as multipole moments of varying order, but the natural way of thinking of the source of a magnetic field is in terms of loops of current.

The set of radial dipoles is related to the reversals of the main field described in chapter 8, for it has recently been found that, during a reversal of the field, the main dipole probably decays to zero and then grows again in the opposite direction, while the subsidiary dipoles, if the residual field is so interpreted, remain apparently unchanged.

9.4 The secular variation

The fact that the declination undergoes slow changes has been known since the middle of the seventeenth century when H. Gellibrand noticed that the declination at London had changed progressively. Figure 9.12 shows how the dip and declination at London have changed from 1576 onwards—the early results, especially for dip, are not of course very reliable, but the general nature of the changes is thought to be well established. The rate of change of any element of the field may be plotted as in figure 9.10, which is drawn for the vertical component, and it will be seen that the curves of equal variation form a number of closed loops, or foci.

The secular variation is a complex phenomenon, in which at least five components can be isolated at the present time—a decrease in the main dipole field by 5 parts in 10^4 per year; a westerly rotation of the dipole at about $0°05$ in longitude per year; a decrease in the angle between the dipole axis and the polar axis of about $0°02$ per year; a drift of the non-dipole field to the west at some $0°2$ per year; and, finally, growth of some features of the non-dipole field and decay of others leading to changes of some 10^{-8} T per year. The drift of both the dipole and non-dipole field to the west is well established; there is some ambiguity in how much of the westerly drift should be assigned to the main dipole and how much to the non-dipole part of the field. Halley was the first (in 1692) to suggest that the secular variation corresponded to a westerly drift; he speculated that the outer part of the Earth, equivalent to the mantle in present-day terminology, might turn eastwards about the inner part, or core, in which the field originated. The modern approach to the study of the secular variation was initiated by Bullard, who showed that the drift of the non-dipole field could be accounted for, if that part of the field were produced by currents in closed loops in the core, those loops drifting to the west under the action of hydrodynamic forces in the core or acting between the core and the mantle. Recent ideas about the generation of the main field by some sort of dynamo action in the core spring from Bullard's investigation of the secular variation.

At the present time it seems that the ratio of the non-dipole to the dipole part of the field is growing rapidly, even though there does not seem to have been a large change in the intensity of the dipole part, which has decreased slowly over the past 200 y, although, because of the difficulty of making absolute measure-

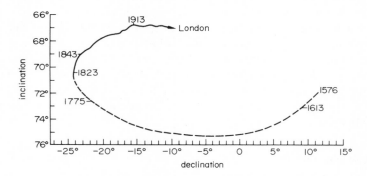

Figure 9.12 Behaviour of inclination and declination at London

ments of the field until quite recently, the change in the intensity is not so well established as those in dip and declination. Over a longer period of time, however, measurements of the intensity of magnetization of lavas and pots that can be dated by archaeological evidence have shown that the intensity has decreased over a period of more than two thousand years. Figure 9.13 shows results obtained in

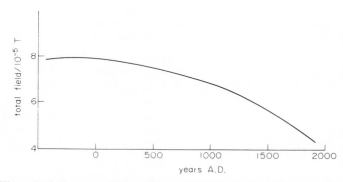

Figure 9.13 Recent variation of the total field in Europe and North Africa

Western Europe and North Africa, indicating that the total field has decreased from over 8×10^{-5} T at the start of the Christian era to about 5×10^{-5} T now; studies in the Caucasus which agree with those in Western Europe for the past 2000 y suggest that in earlier times the field increased and reached a maximum about 2000 y ago.

9.5 The source of the main geomagnetic field

The first point to be made about the source of the main field is that it lies within the Earth. There are indeed fields which vary rapidly, in the course of a day, more or less, and which are produced by currents circulating outside the Earth in the ionosphere, but those fields are relatively small and have no bearing on the origin of the main field which it may be shown unambiguously is produced by sources inside the Earth. The argument depends on the relation between the field components which follows from the fact that the radial factor in the geopotential is $(a/r)^{n+1}$ and not $(r/a)^n$, and is explained in more detail in the next section where it is shown how the rapidly varying components of the field may be separated into parts of internal and external origin. It has also been shown that the sources of the field do not lie within the outermost parts of the Earth, a possibility that had been suggested by the hypothesis of Lord Blackett that magnetization might be a fundamental property of rotating bodies. In that case, all the matter of the Earth would have a magnetization, including the outermost parts, and if measurement were made at depth it would be possible to distinguish the local magnetization from that arising at greater depths. Careful measurements failed to show any local magnetization. Since the measurements extended to a depth of no more than some 1000 m, they do not exclude the possibility that the main field might originate within the lower part of the crust or within the mantle, but consideration of the rates of change of the main field strongly suggests that the cause should be sought below the mantle.

The slow changes of the main field occur with a time constant of some thousands of years, and that major reversals may take place within 10^4 y and last for some 10^5 or 10^6 y. Such times are long compared with the characteristic times of changes associated with the fluid cover of the Earth which, mainly controlled by the Sun and the Moon, are of the order of a day or a year. Going deeper into the Earth, into the crust and mantle, the material is found to be rigid over short time intervals, and processes with time constants of 10^3 or 10^4 y seem most implausible in those zones of the Earth.

It seems clear that the origin of the main field and the secular variation field must be sought in currents within the Earth, for the Curie points of the known ferrimagnetic constituents of the materials within the mantle are exceeded by the temperature within the Earth at very moderate depths, of the order of 20–30 km. It might be argued that the main field was due to ferromagnetic material in the upper parts of the crust, and that the secular variation might be caused by changes in temperature within the crust, but the thermal time constants are too long and the intensity of magnetization would be higher than anything observed at the surface for all but very exceptional rocks. The origin of the field must therefore be looked for in the fluid core of the Earth, within which the time constants of changes of fluid motions are short enough to allow motions with the speed of the westerly drift of the secular variation.

At the present time, the main field is thought to be generated by dynamo action within the core, that is to say, the motions of the conducting liquid core relative to the magnetic field in the core generate currents which produce the magnetic field necessary to maintain them. Evidently, if such a scheme is to work, the electrical conductivity of the core must be great enough, the core must be liquid and the viscosity of the liquid must not be too great. It is usually supposed that the material of the core satisfies these conditions. There must also be a source of energy to drive the movement of the liquid against the viscous and electromagnetic forces that resist it. J. Larmor (1919) first suggested that the magnetic field of a large liquid body could be maintained by dynamo action but T. G. Cowling showed that a very large class of motions could produce no field, namely those which were symmetrical about an axis of rotation. It seemed for a long time to have been thought that Cowling's result ruled out the possibility of a dynamo mechanism for the origin of the magnetic field of the Earth, but E. C. Bullard and W. M. Elsasser argued that it should be possible to have a system of motions in the core of the Earth with sufficient assymmetry to give a self-maintained field.

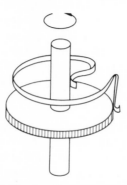

Figure 9.14 Simple self-excited disc dynamo

The disc dynamo will show the principles involved in the dynamo theory. The arrangement shown in figure 9.14 comprises a disc attached to an axle and in electrical contact with a coil that is concentric with the axle. Suppose that there is a magnetic field parallel to the axle. The rotation of the disc in that field will generate an electromotive force directed along the radius of the disc and so drive a current through the coil which, if it is in the correct direction, will produce a magnetic field in the same sense as the original one and will therefore amplify any original field. Two equations govern the behaviour of the dynamo, an electrical equation and a mechanical equation.

If M is the mutual inductance of the disc and the coil, the electromotive force induced across the disc is $\omega M j / 2\pi$, where ω is the angular velocity of the disc and j is the current through the coil. The potential drop in the circuit of the disc and coil is

$$L \frac{\mathrm{d}j}{\mathrm{d}t} + Rj$$

where L is the inductance of the circuit and R the resistance.
Thus

$$L\frac{\mathrm{d}j}{\mathrm{d}t} + Rj = \frac{1}{2\pi}\omega Mj \qquad (9.9)$$

The mechanical equation expresses the balance between the angular acceleration of the disc and the applied torques. Let the mechanical torque applied to drive the disc be G; the electrical torque due to the interaction of the disc with the magnetic field is $Mj^2/2\pi$, and so the net torque accelerating the disc is $G - Mi^2/2\pi$; if I is the moment of inertia of the disc,

$$I\frac{\mathrm{d}\omega}{\mathrm{d}t} = G - \frac{Mj^2}{2\pi} \qquad (9.10)$$

In a steady state where the field is neither increasing nor decreasing, $\mathrm{d}j/\mathrm{d}t$ and $\mathrm{d}\omega/\mathrm{d}t$ are both zero, and so

$$\omega = 2\pi R/M$$

and

$$j = (2\pi G/M)^{1/2} \qquad (9.11)$$

Consider a disc of radius about 1 m and resistance about 0·1 ohm. The mutual inductance of a coil of about the same radius will be about $4\pi \times 10^{-7}$ H and so the angular velocity for the steady state will be about 5×10^6 rad s^{-1}. This is a very high speed and it is unlikely that the simple disc dynamo can be realized in a laboratory, but the calculation shows how in principle a self-acting dynamo may be constructed without any external field. It also shows that the speed needed to maintain the field is inversely proportional to the radius of the disc, so that it is likely that a dynamo the size of the core of the Earth would work at quite slow speeds.*

A dynamo scheme will operate in the core of the Earth only if a suitable set of fluid motions and magnetic fields can exist in a fluid sphere without internal boundaries and if there is an adequate driving force to overcome the dissipation in the viscosity of the liquid and in the Joule heating of the liquid by the electrical currents. The problem has been approached both mathematically and through experimental models.

Mathematically the problem is governed by Maxwell's equations which relate the magnetic field to the electrical currents, and by the equation of a viscous fluid (the Navier–Stokes equation) to which are added terms representing the force arising from the interaction of a magnetic field with an electrical current and also that representing the force on a moving liquid which is rotating about an axis as in the Earth. One consequence of the equations is that, if the conductivity is

* Practical engineering dynamos have complex geometrical arrangements of the coils, and cores of material of very high permeability so that they may operate at very much lower speeds that the simple disc.

sufficiently high, the magnetic field lines move with the fluid—the field is frozen in to the fluid—and then it is possible for rotations of the fluid to twist up the field lines and so amplify the field, as indicated in figure 9.15. In such ways, the magnetic field gains energy from the fluid motions and the role of the electrical currents is to keep the magnetic field frozen in. Mathematical studies have in fact demonstrated that suitable patterns of motions can be found that will maintain a magnetic field.

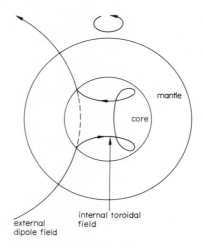

Figure 9.15 Twisting of field lines in a dynamo scheme

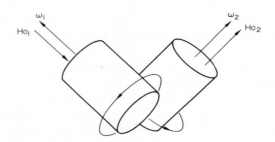

Figure 9.16 The Lowes–Herzberg dynamo

In the experimental approach the aim is to construct a self-maintaining dynamo without the complex paths involving insulation that are necessary in engineering. A dynamo has in fact been constructed of two rotating cylinders with their axes at right angles placed in a bath of mercury (figure 9.16). It was found that, if one cylinder was rotated at a constant speed while the speed of the other was increased, then the external magnetic field suddenly increased at a critical value of the speed of the second cylinder, indicating the onset of dynamo action (figure 9.17).

Closely related to experimental models are calculations done on a computer to simulate the disc dynamo which show that if two such dynamos are coupled together so that the disc of one feeds the coil of the other—an arrangement of greater complexity than the single disc—then the field undergoes spontaneous reversals as indicated in figure 9.18. The pattern is evidently very similar to that of the reversals of the Earth's field and adds support to the idea that dynamo action is in some way responsible for the main field.

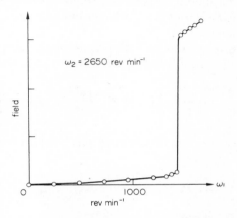

Figure 9.17 Critical behaviour of Lowes–Herzberg dynamo

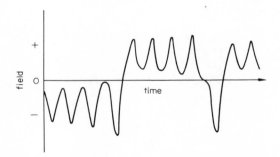

Figure 9.18 Reversals in a computer model of a dynamo

9.6 Electromagnetic induction and conductivity within the Earth

The part of the magnetic field observed at the surface of the Earth that goes through a complete cycle in the course of a day and which is generated by the currents flowing in the ionized gas around the Earth is not confined to the exterior of the Earth but extends some way into the interior, dying off as the depth increases. The material of the Earth does conduct electricity to a slight extent and therefore induced electrical currents flow below the surface generating a magnetic field that does not vanish at the surface and which, like the external field, undergoes a complete cycle in a day.

The primary external field has spherical harmonic components that vary as $(r/a)^n$, whilst those of the secondary internal field vary as $(a/r)^{n+1}$. As will be shown in more detail below, the two fields can be estimated separately by using the different dependence on radius, and so the secondary field can be compared with the primary field. Then, using Maxwell's equations for the electrical and magnetic fields in the interior of the Earth, estimates can be made of the electrical conductivity of the interior of the Earth (essentially in the mantle) that will give the observed relation.

To see how the primary and secondary fields may be separated, let us suppose that the daily variation field is represented by just two harmonic terms:

$$V = \left(\frac{a^4}{r^3} g_{2i} + \frac{r^2}{a} g_{2e} \right) P_2(\cos\theta) + \left(\frac{a^5}{r^4} g_{3i} + \frac{r^3}{a^2} g_{3e} \right) P_3(\cos\theta) \qquad (9.12)$$

where g_{2i}, g_{2e} and g_{3i}, g_{3e} are functions of time having a period of one day.

Then

$$Z = -\frac{\partial V}{\partial r}\bigg|_{r=a} = (3g_{2i} - 2g_{2e}) P_2(\cos\theta) + (4g_{3i} - 3g_{3e}) P_3(\cos\theta) \qquad (9.13)$$

while Y is zero because it is assumed that the potential does not depend on longitude.

Now

$$P_2(\cos\theta) = \tfrac{1}{2}(3\cos^2\theta - 1)$$

and

$$P_3(\cos\theta) = \tfrac{1}{2}(\cos^3\theta - 3\cos\theta)$$

and so

$$\frac{\partial P_2}{\partial\theta} = -3\sin\theta\cos\theta$$

and

$$\frac{\partial P_3}{\partial\theta} = -3(\tfrac{1}{2}\cos^2\theta - 1)\sin\theta$$

and thus

$$H = -\frac{\partial V}{r\,\partial\theta}\bigg|_{r=a} = 3(g_{2i} + g_{2e})\sin\theta\cos\theta$$

$$+ 3(g_{3i} + g_{3e})(\tfrac{1}{2}\cos^2\theta - 1)\sin\theta \qquad (9.14)$$

The expressions for Z and H may be written as

and

$$\left.\begin{array}{l} Z = z_2 P_2(\cos\theta) + z_3 P_3(\cos\theta) \\[2mm] H = 3h_2 \sin\theta\cos\theta + 3h_3(\tfrac{1}{2}\cos^2\theta - 1)\sin\theta \end{array}\right\} \qquad (9.15)$$

where

$$\left.\begin{array}{l} z_2 = 3g_{2i} - 2g_{2e} \\[1mm] z_3 = 4g_{3i} - 3g_{3e} \\[1mm] h_2 = g_{2i} + g_{2e} \\[1mm] h_3 = g_{3i} + g_{3e} \end{array}\right\} \qquad (9.16)$$

Numerical analysis of the daily variation fields gives the quantities z_2, z_3, h_2, h_3 and the internal and external coefficients then follow from

$$\left.\begin{array}{l} g_{2i} = \tfrac{1}{5}(z_2 + 2h_2) \\[1mm] g_{2e} = \tfrac{1}{5}(3h_2 - z_2) \\[1mm] g_{3i} = \tfrac{1}{7}(z_3 + 3h_3) \\[1mm] g_{3e} = \tfrac{1}{7}(4h_3 - z_3) \end{array}\right\} \qquad (9.17)$$

The analysis can be applied to more complex fields, and an example of the results is shown in figure 9.19. The separation can also be carried out by using certain properties of the integrals of the field over the surface of the Earth rather than by making a spherical harmonic analysis of the variation fields.

The analysis must be performed for each separate variation in time, of which there are two main types. There is the storm-time variation (Dst) in which the field undergoes a sudden change (ssc) that decays away to zero in some two days, and of which the distribution over the surface of the Earth can be represented by three zonal harmonics, $P_1(\cos\theta)$, $P_3(\cos\theta)$ and $P_5(\cos\theta)$. Figure 9.19 shows the analysis of the storm-variation field. The storm-time variation is superposed on the quiet-day cyclic variations of the field that can be represented by harmonics of the type $P_n^{n-1}(\cos\theta)$ or $P_n^n(\cos\theta)$, the most important being those for which n is 1, 2, 3, and 4.

The currents induced in the Earth by the external variation fields satisfy Maxwell's equations, as do the fields that they generate outside the Earth. If the electrical field is E, it may be shown that E satisfies

$$\nabla \wedge \nabla \wedge E = -\mu\sigma\, \partial E/\partial t \qquad (9.18)$$

μ is the magnetic permeability and σ is the electrical conductivity, both of which in general vary with radius r. However, μ varies much less than σ and so it is sufficient to take it to be a constant. The normal component of E must vanish at the surface $r = a$, and the equation can then be satisfied by

$$E = r \wedge \nabla u$$

where u is a function that satisfies

$$\nabla^2 u = -\mu\sigma(r)\,\dot{u}$$

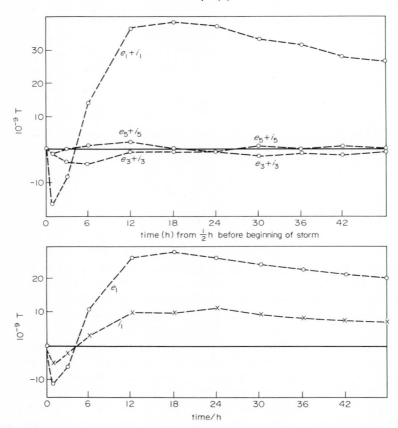

Figure 9.19 Separation of harmonic components of storm-time variation field (Dst)

For this solution

$$\nabla \cdot E = 0$$

and so there is no volume distribution of electrical charge.

If u is written as the product of a surface harmonic S_n and a time-varying radial function $R_n(r)$, it follows that

$$\frac{\partial}{\partial r}\left(r^2 \frac{\partial R_n}{\partial r}\right) = \left\{n(n+1) - \mu\sigma r^2 \frac{\partial}{\partial t}\right\} R_n \qquad (9.19)$$

R_n can be calculated, usually numerically, for a given variation of σ with depth and a given frequency, and the value at the surface may then be expressed as the sum of parts proportional to $(r/a)^n$ and $(a/r)^{n+1}$. The former represents the externally generated part of the field, the latter the internally generated part. Since both are known, it can be seen whether the postulated variation of σ with radius leads to agreement with the observed fields.

In the first calculations by S. Chapman and A. T. Price, it was supposed that the Earth could be divided into two zones, a thin outer layer of low conductivity and the remainder of the Earth with high conductivity. If only the daily variation is used in the investigation, or only the storm-time variation, then the data can be satisfied by a very wide range of distributions of conductivity within the Earth, but the range can be narrowed down somewhat by using both variations. It remains true, however, that a unique distribution of conductivity cannot be obtained, and figure 9.20 shows a number of models that have been worked out. The faster the variations of the surface field, the less deep do the currents that they induce penetrate into the Earth (the skin depth effect in an electrical conductor) and so the conductivity at the greater depths cannot be found from the daily or storm-time variations but must be inferred from the secular variation. As shown in figure 9.20, the conductivity rises from about 10^{-2} mho m^{-1} at the surface to about 100 mho m^{-1} at the base of the mantle; for comparison, the conductivity of copper is 5×10^7 mho m^{-1} at room temperature, that of germanium is about 1000 mho m^{-1} and that of distilled water ranges from 10^{-2} to 10^{-5} mho m^{-1}.

The range of conductivities found in the mantle is between that of good conductors (10^7 mho m^{-1}) and of good insulators and is similar to that of semi-conductors. In a metal, a wide range of states of energy and momentum is available to electrons which, in consequence, would move freely through the crystal lattice of a metal were it not for collisions with the lattice changing the momentum and leading to a resistance to motion. In semiconductors the states available to the electrons fall into two groups—the 'valence band' in which all states are normally occupied, and the 'conduction band' in which only few electrons occupy the available states (see, for example, Kittel, 1968). Electrons in the valence band cannot contribute to the transport of electricity because transport entails electrons changing their momenta and, if all the available states are occupied, no electrons can change from one state to another so that no changes of momenta occur and no transport of electricity occurs. A material with all the available states always occupied would be a perfect insulator. In a semiconductor, another group of states—the conduction band—can be occupied but there is a difference between the highest energy of the valence band and the lowest of the conduction band (figure 9.21) which is comparable with the Boltzmann energy, kT, for room temperature, so that very few electrons are in the conduction band. Electrons in the conduction band have a range of unoccupied states open to them and so can transport charge under the influence of an electric field. The conductivity is proportional to the number of electrons in the conduction band which is in turn proportional to $\exp(-E_g/kT)$, where E_g is the difference of energy between the top

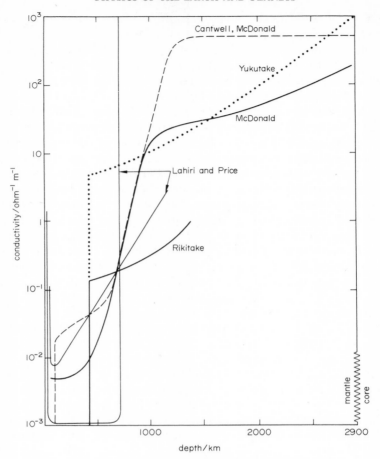

Figure 9.20 Possible variations of conductivity within the Earth

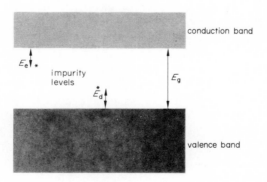

Figure 9.21 Energies in a semiconductor

of the valence band and the bottom of the conduction band. Thus the conductivity is given by

$$\sigma_{i0} \exp(-E_g/kT)$$

where σ_{i0} is a quantity dependent on the lattice structure of the crystal and is thus a function of lattice spacing and therefore of pressure.

In addition to this essential or *intrinsic* conductivity, impurities may give rise to additional transport of electricity, either by providing electrons which may be raised into the conduction band (donor impurities) or by receiving electrons from the valence band (acceptor impurities). In the latter case removal of electrons from the valence band makes it possible for those remaining to change their state when an electric field is applied and so for transport of electricity to take place. The conductivity due to impurities is known as extrinsic conductivity and is proportional to

$$\sigma_{e0} \exp(-E_e/kT)$$

E_e is the difference of energy between the electrons associated with the impurities and the conduction band when the impurity is a donor, and between the energy of the impurity electrons and the top of the valence band when the impurity is an acceptor. σ_{e0} now depends on the concentration of impurities as well as on the lattice structure.

Lastly, it is possible for the ions to detach themselves from a particular site and to diffuse through the lattice when an electrical field is applied. In that case (ionic conductivity) the conductivity is given by

$$\sigma_{z0} \exp(-E_d/kT)$$

where E_d is the energy an ion must acquire for it to become detached from a lattice site.

The variation of electrical conductivity within the Earth is controlled both by the variation of temperature, which enters the Boltzmann factor, E/kT, directly, and through pressure, which changes the value of the excitation energy through change in lattice spacing of the crystal. Pressure always decreases the mobility of ions by decreasing the lattice spacing (which is of the same order as the size of an ion); thus, as temperature and pressure both increase with depth, the variation of ionic conductivity with depth will depend on a balance between the increase of T and the increase of E_d. Calculations suggest that the ionic conductivity of olivine will reach a maximum somewhere within the upper mantle and will then fall off. The same calculations also indicate that the conductivity of the upper part of the upper mantle is mainly ionic but, since the conductivity of the mantle rises sharply with depth instead of staying more or less constant, it follows that the electronic conductivity mechanisms must also come into effect at greater depths.

The gap between the conduction and the valence bands may either increase with pressure or decrease, depending on lattice structure. The gap may in fact disappear through overlap of the two bands, in which case the material behaves as a metal; transitions from the non-metallic state to a metallic form under pressure are indeed well known. On the other hand, the resistivity of some semiconductors increases with pressure. Measurements on olivine have shown that the intrinsic gap decreases with pressure at a rate given by

$$\frac{1}{E_g}\frac{dE_g}{dp} = -6 \times 10^{-12} \ (p \text{ in N m}^{-2})$$

Thus, since E_g decreases with depth and T increases, it is to be expected that the intrinsic conductivity of olivine will increase sharply with increasing depth, corresponding to what is found in the Earth.

It was explained in chapter 7 that the transport of heat in a non-metallic solid may occur through three processes: lattice vibrations, radiative transfer and exciton transfer. Transfer of heat within a metal, on the other hand, is by the free electrons. The transfer of electricity within a metal, being also by the free electrons, is related to the transfer of heat by the Wiedemann–Franz law:

$$\kappa = 3\left(\frac{k}{e}\right)^2 \sigma T$$

where κ is the thermal conductivity, σ the electrical conductivity, k is Boltzmann's constant, e is the electronic charge and T the temperature. Thus in the core of the Earth, if σ is about 100 mho m^{-1}, as is suggested by a dynamo theory of the main magnetic field, κ will be about 10^{-4} W m^{-1} degK^{-1}. Unfortunately there is no corresponding relation for insulators and semiconductors since electrons play a minor part in heat transport. The main interest of the estimates of electrical conductivity within the Earth is therefore in the indication given by the sharp increase at about 500 or 600 km that the variation of temperature derived in chapter 7 is generally of the correct order, and in the possibility that the electrical conductivity within the Moon and the planets may give some idea of the corresponding temperatures (chapter 11).

The sharp increase of electrical conductivity occurs in the region where the polymorphic change in the structure of olivine is thought to take place. That change will also affect the conductivity because, as mentioned before, the value of E_g depends on the lattice structure. Experiments show that the spinel form of olivine has a conductivity about 100 times greater than that of the low-pressure olivine; the observed increase of conductivity in the region of 500 to 800 km is more nearly by a factor of 10^4 and so the effect of increase of temperature must also be invoked to account for the data.

Examples for chapter 9

9.1 A vertical field magnetometer consists of a rubidium vapour magnetometer surrounded by a coil generating a vertical field (see p. 213). Let the field generated by the auxiliary coil be in one case 5×10^{-6} T upwards and in the second case 5×10^{-6} T downwards, the corresponding frequencies of oscillation of the magnetometer being 149·4 kHz and 188·2 kHz.

Calculate the vertical field, the total field and the inclination.

The instrumental constant for the rubidium magnetometer is $4·67 \times 10^9$ Hz T^{-1}.

9.2 Solve equation (9.19) under the following condition:
$$u = S_1(\theta, \phi) R_1(r) e^{i\omega t}$$
and
$$\sigma = \sigma_0/r^2$$
Attempt a solution of the form $R_1(r) = r^p$.

9.3 Calculate the depth of penetration, δ, of currents within the Earth for periods of 1 d and 1 h. Take δ to be given by
$$(2/\omega\mu\sigma)^{1/2}$$
and assume
$$\mu = 1·3 \times 10^{-6} \text{ H m}^{-1}$$
$$\sigma = 1 \text{ mho m}^{-1}$$

9.4 Show that the total field of a dipole at the Earth's surface is proportional to
$$\{2 + 2p_2(\cos \theta)\}^{1/2}$$
where θ is the colatitude.

Compare the behaviour of this expression with that shown by the Earth's field in figure 9.9.

Further reading for chapter 9

Alldredge, L. R., and Hurwitz, L., 1964, 'Radial dipoles as the source of the Earth's main magnetic field', *J. Geophys. Res.*, **29**, 2631.

Chapman, S., and Bartels, J., 1940, *Geomagnetism*, Vols. 1 and 2 (Oxford: Clarendon Press).

Jensen, D. C., and Cain, J. C., 1962, 'An interim geomagnetic field', *J. Geophys. Res.*, **67**, 3568.

Kittel, C., 1968, *Introduction to Solid State Physics*, 3rd edn (New York, London, Sydney: Wiley), 648 pp.

CHAPTER TEN

The structure and movements of the crust of the Earth

'Stands Scotland where it stood?'
Shakespeare, *Macbeth*, IV, 3

10.1 The structure of the crust

By following the methods described in chapters 2 and 3, we have learnt a great deal about the structure of the crust of the Earth down to the top of the upper mantle. The principal result is that there is a clear distinction between the crust under the continents and that under the oceans, as shown in the summary sections of figure 10.1. The continental crust is upwards of 30 km thick and composed of material akin to the acidic and intermediate rocks seen at the surface of the continents, while the crust under the oceans is much thinner and composed of distinct layers, namely uncompacted sediment, a layer that may be igneous or compacted sediment and a layer with properties corresponding to basalt at the surface. As shown in table 3.2, the densities and thicknesses of the various layers ensure that there is isostatic balance between the continents and oceans. The summary given in figure 10.1, while generally true, is too simple, at least for the continents, although the oceanic crust is for the most part rather uniform.

In the first place, the crust is sometimes divided into two layers by a boundary known as the *Conrad* discontinuity, evidence for which was first discovered in 1925. The Conrad discontinuity is not always clear in seismic refraction studies and it seems that it is a true boundary in some places but not in others. Russian investigations of the crust and upper mantle by reflections of seismic waves (deep seismic sounding) have found a clear Conrad discontinuity in western Russia and south-eastern Europe, as shown in figure 10.2. If the Conrad discontinuity occurs it appears to separate an upper crust with a P-wave velocity of $5 \cdot 8$–$6 \cdot 4$ km s^{-1} and a density of 2700–2800 kg m^{-3} from a lower crust with a P-wave velocity of $6 \cdot 6$–$7 \cdot 0$ km s^{-1} and a density of about 3000 kg m^{-3}.

In other areas, lateral variations in the continental crust and mantle are more apparent than vertical ones. Figure 3.26 shows the variation of the thickness and seismic velocity of the crust in the continental U.S.A., and the variation in the seismic velocity of the mantle immediately below the crust.

The crust used to be spoken of as made of granite, partly because it seemed that the density of the crust might be similar to that of granite, partly because of the very large masses of granite that are seen at the surface, especially in mountains like the Alps or the Rocky Mountains, and partly because the prevalence of rocks made of sand at the surface suggests that there was a parent rock with a lot of silica in it, namely granite. It now turns out that the acceleration due to gravity is less over almost all large masses of granite than over the surrounding rocks, implying that the density of granite is less than that of the typical material of the

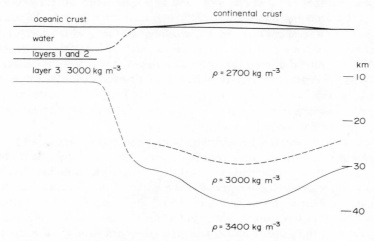

Figure 10.1 Schematic diagrams of oceanic and continental crust

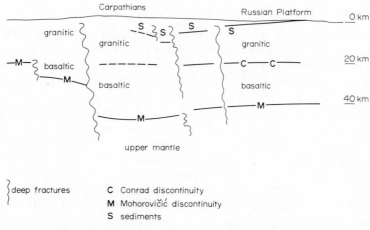

Figure 10.2 The crust of the south-west of the U.S.S.R.

crust near the surface. The sizes of the gravity anomalies show that the largest masses of granite must extend to a depth of more than 15 km, and it seems clear that the upper part of the crust is composed for the most part of material that is somewhat denser than granite.

The way in which oceanic crust changes to continental crust appears to depend in part on whether the Conrad discontinuity can be distinguished in the continent. In figure 3.25, for example, the lower continental crust has approximately the same thickness and density as the lower layer of the oceanic crust and is continuous with it, while the upper layer thins out at the margin. Rivers on the continents bring down to the seas quantities of sand and mud worn off the rocks of the continents by wind and rain, by sun and frost, and deposit them at the edge of the continents forming a bank called the continental shelf, which merges on the seaward side into the sedimentary layer of the oceans. On some coasts the sediments of the continental shelf are no thicker than the depth of the ocean, but on other coasts the crust is depressed under the continental shelf and attains a great thickness. When the Conrad discontinuity cannot be distinguished on the continents, the whole continental crust then thins down and the density increases as it merges into the lower oceanic crustal layer.

The distinction between the continents and the oceans appears to be quite essential, but it has already been seen in chapter 2 that isostatic balance between the continents and the oceans is so good that on a world-wide basis there is no indication in the general gravity field of the Earth of the presence of continents and oceans. Similarly, it was seen in chapter 7 that the flow of heat through the continents is almost identical with that through the oceans. There are thus indications that the distinction between the continents and the oceans may be less fundamental than appears at first sight.

10.2 Movements of the crust

The crust of the Earth is in continual movement. Parts of it flow out molten upon the surface as lavas from volcanoes, it breaks in earthquakes and sections move relative to each other, and changes in sea level show that it rises and falls. These things may be seen from day to day or perhaps in the course of the life of a man, but many features of the rocks show that larger movements have occurred in the past over millions or tens of millions of years. Faults are surfaces along which rocks have moved against each other, sometimes up to more than a hundred kilometres, and the rocks in mountains can be seen to have been deformed to an extreme degree, while the very presence of mountains themselves, composed as they are of rocks that once lay at great depth upon the sea bed, witness to movements that have taken place over hundreds of kilometres and raised rocks through tens of kilometres. Such plain witness to the movements of the crust lead us to ask whether there may be a pattern to be discerned in them, and whether we can investigate the forces that drive them and sources of energy by which they are fed.

The study of movements of the crust has advanced considerably within the last decade as the result of new methods of measurement, in particular, the direct measurement of movements at the surface and the study of the permanent magnetization of rocks. It was seen in chapter 8 that many old rocks carry a permanent magnetization that indicates that the continents on which they were formed have moved relative to each other. For example, the Atlantic Ocean appears to have opened up in the last 200 My, an excursion of some 3000 km on each side, or a rate of movement of about 1·5 cm y⁻¹. It would be quite wrong on such evidence alone to argue that movements were going on at the present time at the same rate, but other observations suggest that the average rate over much more recent times is similar.

The median line of the North and South Atlantic Oceans is marked by a ridge that rises some 3 km above the sea floor on either side and that is over 1000 km in width. It is therefore a much larger feature of the crust of the Earth than even the greatest mountain ranges. The sides of the ridge are very rugged unlike the sea

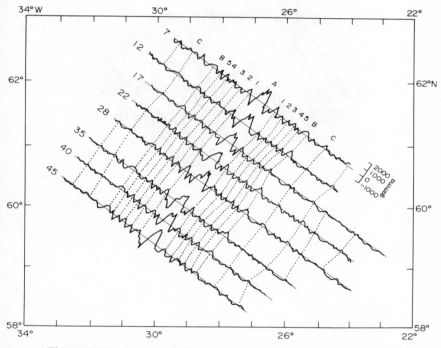

Figure 10.3 Magnetic traverses south-west of Iceland (see Bullard, 1968)

bed on the flanks (the abyssal plain) which is rather smooth, and the middle line of the ridge is marked by a discontinuous deep valley with steep sides. A typical section through the mid-Atlantic ridge is shown in figure 3.28. Similar ridges are found in all the oceans, though usually not so symmetrically disposed as in the Atlantic ocean; a map of the world-wide ridge system is shown in figure 3.27. Gravity anomalies over the ridges are usually low and seismic investigations have shown the velocity of seismic waves below the ridges to be low also, indicating that the upper mantle below the ridges is less dense than elsewhere. Figure 3.29 gives a general idea of the structure below a ridge.

Measurements of the magnetization of the rocks of the ocean floor made with magnetometers towed behind aircraft have shown a very simple pattern over the median ridges of the oceans. One of the earliest and still one of the clearest examples is the pattern over the mid-Atlantic ridge to the south-west of Iceland. The magnetic field that is observed at the height of an aircraft is the resultant of the field induced by the present field of the Earth together with the field due to the permanent magnetization of the rocks of the oceanic crust down to the depth at

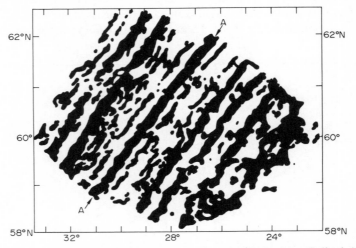

Figure 10.4 The magnetic stripe pattern south-west of Iceland (see Bullard, 1968)

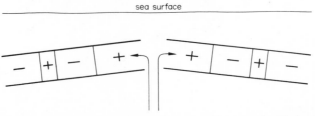

Figure 10.5 Sea floor spreading and magnetization: +, rocks magnetized in present direction of Earth's field; −, rocks with reversed magnetization

which the temperature exceeds the Curie temperature. The measurements are made in parallel lines and some of the traverses are shown in figure 10.3. When the magnetization induced by the present field is subtracted, it is found that the permanent magnetization is arranged in parallel strips, strips magnetized in the same direction as the present field alternating with those magnetized in the opposite direction, as shown in figure 10.4. Here, as in other parts of the world, the pattern is found to be disposed symmetrically about the central valley (AA) of the associated mid-oceanic ridge. A very plausible explanation of the structure was given by F. A. Vine and D. B. Matthews (1963); they suggested that volcanic rock is continually rising along a mid-oceanic ridge, as is indeed indicated by the presence of submarine volcanoes along ridges, and that, as the rock cools through the Curie point, it becomes permanently magnetized in the prevailing direction of the main magnetic field of the Earth, at times in the same direction as the present field and at times in the opposite direction. As more material rises from the mantle to the ocean floor, the material already present is forced sideways to accommodate

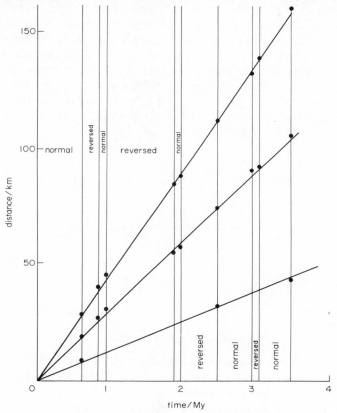

Figure 10.6 Rates of sea floor spreading

the new material and so, as indicated in figure 10.5, the magnetization of the rocks of the sea floor is arranged in strips parallel to the mid-oceanic ridge from which the rocks originated, the direction of magnetization corresponding to the direction of the main field of the Earth at the time when the particular section of the floor cooled below its Curie point, or, near enough, the time when it rose into the crust. The magnetization of the rocks of the mid-oceanic ridges is therefore seen as closely connected with the separation of the continents.

Now it was seen in chapter 8 that the times at which the main field of the Earth have changed direction are quite well known, and therefore, by identifying the episode of normal or reversed magnetization during which a strip of the sea floor was formed, it is possible to say how long ago it was formed and therefore to estimate the rate at which it moved outwards from the mid-oceanic ridge between that time and now. Figure 10.6 shows distances plotted against times for some ridges, and it will be seen that the rate of movement lies between 1 and 6 cm y^{-1} over the past 5 My, rates that correspond very closely with those derived from the presumed separation of the continents over the much longer interval of 200 My.

The topography of the mid-oceanic ridges as worked out from measurements of depths with echo sounders shows that the ridges are intersected by lines almost at right angles to the median valley, the valley itself often being displaced parallel to itself at such a line. It is now thought that the lines divide the sea floor into blocks that move at different times or rates, as indicated in figure 10.7. In the upper part of the diagram, the mid-oceanic ridge divides block A from block A', while in the lower part of the diagram the ridge is displaced to the right and separates block B from block B'. Evidently, as blocks of a pair separate, there is

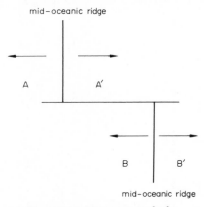

Figure 10.7 Transcurrent faults

little relative movement between A and B or between A' and B' but, in the centre region, block A' which is moving to the right lies against block B which is moving to the left, and the relative movement between these two is twice the rate at which any block is moving outwards from the mid-oceanic ridge. Perhaps then earthquakes should be especially frequent along these lines between offset median ridges and a detailed study of the many earthquakes that occur along the mid-oceanic ridges shows that they do indeed lie for the most part in the expected places.

The movement along a fault can in favourable circumstances be found from the records of the signals that arrive at stations all over the Earth. Consider the simplest possible movement, that following the explosion of an atomic bomb buried in the Earth. The first thing that happens is that the rocks are compressed symmetrically around the bomb and the compression is then followed by a similarly symmetrical rarefaction. The first signal to arrive at a seismometer anywhere over the Earth would therefore be expected to correspond to a compression of the ground at the seismometer and would be followed by a signal corresponding to rarefaction. Now consider the simple sliding of one section of rock against another horizontally along a vertical plane. Such a simple shearing motion may be represented by compression in one part of the horizontal plane combined with rarefaction in the other part, as indicated in figure 10.8, and so the surface of the Earth may be divided by great circles into two regions that will see compression at the first movement on a seismic record, and two other regions which will see rarefactions as the first movement. The direction of the vertical

Figure 10.8 Stress pattern for simple shear

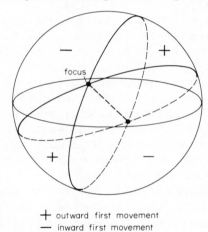

+ outward first movement
− inward first movement

Figure 10.9 Pattern of first motions for simple shear

plane and the direction of movement along it may be deduced from the positions of the great circles that separate the four regions on the surface of the Earth (figure 10.9). Such first motion studies, as they are called, show that movement in faults at right angles to the mid-oceanic ridges is as indicated in figure 10.7.

Three lines of evidence thus appear to demonstrate that over periods from 200 My ago up to the present time, the sea floors have been continually added to at the mid-oceanic ridges and are spreading out from them at rates of a few centimetres per year. While small, such rates should not be impossible to measure directly, and it is natural to ask if there may be any chance of doing so. The difficulty is that, while a centimetre is a very large distance in terms of the sensitivity of modern methods of measurement, it is a very small fraction of any length that would in practice have to be measured. In the first place, a site must be found where the two sides of the mid-oceanic ridge come to the surface close enough for there to be a direct line of sight between them. Only one place is known where that condition is satisfied, namely Iceland, which lies directly over the mid-Atlantic ridge and which is traversed by faults and lava vents corresponding to it. Within Iceland it is possible to find places where the main movement across the faults is confined to a band of one or two kilometres width; thus to detect a change of one or two centimetres, distances must be measured to better than 10 parts in a million. Lengths of a few kilometres can be measured to about 10 parts per million in terms of the time of travel of a ray of light, the main uncertainty coming from uncertainties in the refractive index of the air along the path. Many instruments have been produced in the last twenty years for measuring distances of the order of 30 km to about one part in 10^5 or 10^6 using the time of travel of a ray of light or a beam of short radio waves but, having been developed for large-scale geodetic survey, they read to two or three centimetres instead of to the millimetre or better needed to detect the movements of the continents with any certainty in a period of two or three years.

An instrument developed at the National Physical Laboratory for engineering measurements has proved to have the precision needed for investigating movements at the mid-Atlantic ridge. The Mekometer is an electro-optical instrument for the measurement of distances of the order of 5 km or less that was developed by Dr K. D. Froome (Froome and Essen, 1969). Light is passed through an optically active crystal which rotates the plane of polarization of the light on the application of an electrical voltage. The light is transmitted to a reflector at a distant point and returned to the optically active crystal, through which it will pass unimpeded if the time of travel is such that the plane of polarization is the same on return as it was on transmission. If the voltage applied to the crystal alternates at some 500 MHz, the sensitivity of the instrument is such that a change of 30 μm can be detected in the most favourable conditions. The practical accuracy of the measurement of a length of, say, 2 km is much less because of the difficulty of estimating the refractive index of the air along the path with corresponding accuracy, but nonetheless it has proved possible to make measurements in Iceland that do show movements of the order of 1 to 2 cm over 2 km in the course of $1\frac{1}{2}$ years. It is, however, too early to say whether the movements that have been detected are consistent with the supposed movements away from the mid-Atlantic ridge.

Although of great interest in showing that it is possible in places to measure movements of the order of magnitude of the yearly separation of the continents, there can be only very few places indeed where measurements over so short a distance as some 5 km can be possible. Great interest has therefore been aroused by the potentialities of measurements of distances to the Moon by laser. The astronauts from the American *Apollo 11* moon landing placed upon the Moon a reflector consisting of a number of cube-corner reflectors, which, as will be recalled from chapter 3, have the property that they reflect light back parallel to the incident direction. Thus, when the reflectors are illuminated by a pulse of light from a laser at an observatory on the Earth, a reflected pulse is sent back to the same point on the Earth where it may be received by a telescope of sufficient aperture. If the received light is concentrated by the telescope on to the cathode of a photomultiplier, an electrical pulse is generated and the time of flight of the pulse of light from the laser to the Moon and back may be found from the interval between the laser pulse and the photomultiplier pulse. The pulse should last for little more than 1 ns. In the time since the *Apollo* landings, returned pulses have been received at the Lick and Macdonald observatories in the U.S.A. and, in addition, a second reflector has been placed on the Moon by the Russian Lunokhod expedition and two further ones in later *Apollo* landings.

The theory of the use of the measured distance was given in chapter 4 (figure 4.3). Let R be the distance between the centres of the Earth and the Moon, let r_1 be the distance between the centre of the Earth and the observatory at the surface of the Earth, and similarly let the distance between the centre of the Moon and the reflector on the Moon be r_2. The corresponding vector quantities will be denoted by R, r_1 and r_2. Let D be the vector distance between the observatory on the Earth and the reflector on the Moon, so that the measured distance will be D, the modulus of the vector distance.

Then it was seen that D is very nearly

$$R - p_1 - p_2$$

where p_1 is the projection of r_1 on R and p_2 that of r_2.

Of the various quantities in the expression for the measured distance, R varies with the orbital motion of the Moon around the Earth, while r_1 and r_2 remain constant in magnitude but change in direction relative to R. In particular, because of the spin of the Earth, r_1 rotates about the polar axis once a day and by making use of that fact it will be seen that the p_1 undergoes a daily variation with an amplitude that depends on the latitude of the station and the magnitude of r_1, and with a phase that depends on the longitude of the observatory. Thus from the daily variation of the measured distance D it should be possible to determine the geocentric coordinates of the observatory, and there are indications that in the course of a few months' observations it should be possible to do so to some 10 cm. If that prediction is borne out, then measurements extending over a decade would detect variations of the coordinates of an observatory of the order of a few centimetres in a year, and in particular would be capable of detecting changes of this order in the separation of observatories in different continents.

Horizontal movements are not the only ones to take place and in fact vertical movements are at first sight the more obvious. Some are associated with changes in load on the crust, two examples being the rise of Scandinavia and Finland following the melting of the ice cover of the recent ice age and the similar rise of Lake Bonneville, the extension of the Great Salt Lake, in Utah. The movements are revealed by ancient shore lines that show where the land stood relative to water level. In Scandinavia, the conditions are very clear, for the shore lines show the level of the land with respect to the sea and the times between the various levels have been estimated from varved clays, deposits of clay formed in still glacial lakes with thick and thin bands in an annual cycle. Both in Scandinavia and in Lake Bonneville it is possible to interpret the slow rise of the land after the removal of the load in terms of an elastic crust upon a viscous mantle and to estimate the viscosity as 10^{21} N s m^{-2}.

Other instances of the land surface rising and falling relative to sea level have been observed as the result of accurate measurements of height repeated at intervals of a few years. The movements are in general very slow, perhaps only 0.01 cm y^{-1}, but larger movements are found in volcanic areas, amounting to many centimetres in a year in some cases.

10.3 The pattern of movements at the surface of the Earth

As the previous section indicates, there is now evidence that over periods of 200 My, 5 My and 2 y, the crust of the Earth has moved at rates of the order of between 1 and 5 cm y^{-1}, but it remains to ask whether the patterns of movement over these very different spans of time are similar? The answer turns to a large extent on whether movement occurs only at the mid-oceanic ridges or whether it may not also take place elsewhere.

Such magnetic surveys of the oceans that have been made do suggest that the characteristic stripe pattern is to be found only in association with mid-oceanic ridges, but that is inadequate proof that the patterns will not be found in other regions because, following the discovery of the stripe patterns, surveys have been made mostly in the neighbourhood of ridges and the areas where the occurrence of patterns has been thought unlikely have been neglected. The study of where earthquakes take place provides a better guide to the concentration of movement on the ridges. As shown in figure 3.1, earthquakes are concentrated along the mid-oceanic ridges and also along the belt of mountains stretching from Malaya through the Himalayas to the Alps. It was also seen in section 10.2 that the earthquakes associated with ridges lay along transcurrent faults and had movements predominantly at right angles to the median line of the ridge. The earthquake evidence thus corroborates that from the magnetic stripe patterns that movement takes place outwards from the ridges, but in addition the almost complete absence of earthquakes elsewhere save along the Alpine–Himalayan belt is a clear indication that no significant movement occurs anywhere else at present.

Given the direction of movement from earthquake displacements and the direction and amount from magnetic stripe patterns, it has proved possible to show that large sections of the surface of the Earth move as rigid plates, that is to say, that the movements of one part of a region are consistent with those of another part, assuming only that each corresponds to a rotation of the region as a whole about some common direction—it was first shown by Euler that any movement of a rigid area upon a spherical surface can be considered to be a rotation about some properly chosen direction. It does seem that the major horizontal movements of the surface of the Earth can be considered to be the

relative movements of some six rigid plates, the distinctive feature of this interpretation being that any one plate comprises in general both continental and
oceanic crust, as shown in figure 10.10. The boundaries between plates are for the
most part mid-oceanic ridges but it can be seen from figure 10.10 that they do in
some places run through continental areas—especially in California, East Africa
and the Red Sea and the Alpine–Himalayan line. The Pacific seems to form one
large plate whereas the Indian Ocean and the Atlantic Ocean lie across plate
boundaries.

The present division of the surface of the Earth into plates corresponds in
general to the relative movements of the continents as inferred from the permanent
magnetization of rocks. There were apparently two major continents 200 My ago,
one comprising Asia, Europe, North America and Greenland and the other,
southern one, often known as Gondwannland, incorporating South Africa,
India, Australia, Antarctica and South America. The subsequent movements
include the separation of Europe, North America and Greenland through the
growth of the Atlantic by the formation of new oceanic crust at the median ridge,
and the separation of the southern continents by the similar growth of the Indian
and Antarctic Oceans. The general pattern seems simple enough—the growing
oceans carrying with them their associated continents—and accounts for major
features of the crust of the Earth, in particular the movements that appear to be
taking place at the present time and the dominant tectonic feature of the oceans,
the mid-oceanic ridges. Other problems are left unresolved or are raised by the
hypothesis of spreading from the median lines of the oceans by growth of the
oceanic crust.

The first problem is the geometrical one of the total area of the crust—does it
change as a result of the growth of the oceans and, if not, where is crust destroyed
to keep the total area the same? If the area had changed, it would imply that the
volume of the Earth and therefore either the mass or the density of the Earth had
changed, over the past 200 My. The combined areas of the Atlantic and Indian
Oceans are $1\cdot55 \times 10^{14}$ m^2 while the surface area of the Earth is $5\cdot1 \times 10^{14}$ m^2 so
that, if no crust had been destroyed, the surface area would have increased by 30%
in the past 200 My, implying an increase in mass or a decrease in density of the
Earth of 45%. There is no direct evidence for such changes and such indirect
evidence as exists is inconclusive, so that the geometrical implications of the sea
floor spreading idea should be considered on the basis that the surface area of the
Earth has remained constant. Some way must therefore be found to account for
the disappearance of crust.

It is generally thought that the island arcs are places in which crust is dis-
appearing below the surface. A section through an island arc, such as the West
Indies or the East Indies, was shown in figure 3.26. Two features common to arcs
are important to the present argument, the deep trough usually filled with
sediments and the distribution of earthquakes. Below the trough, the crust is
depressed into the mantle and even though the trough may be filled with sediments,
gravity is low over the trough, implying that the trough is not in isostatic equilib-

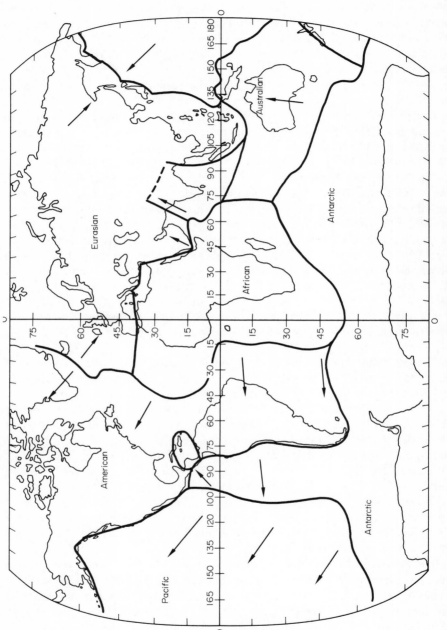

Figure 10.10 Map of plates

rium and that there is some force keeping it depressed. A compressional force tangential to the surface would suffice. Earthquakes associated with island arcs occur along a roughly conical surface that extends at about 45° below the arc. The directions of movements along some of the earthquakes have been determined and are found to be as indicated on the diagram, the motion shortening the crust as would be expected if the crust yields to a compressional stress. There is thus evidence that the crust is under compressional stress in an island arc and that it is yielding to that stress. But it does not follow that crust is being destroyed; it may be that it is just becoming thicker. Indeed that must be happening to some extent, for mountains, where the crust is thickened, appear to arise from island arcs through compression of the crust. The question, which seems open at present, is whether the mountains known to have formed in the past 200 My can provide sufficient thickening of the crust to correspond to the production of (much thinner) oceanic crust.

A question that is related to the possible destruction of crust is the evidence for the growth of continents that was set out at the end of chapter 6 where it was seen that the oldest pre-Cambrian rocks in North America were arranged concentrically, the younger on the outside. A somewhat similar pattern occurs in the more recent rocks, the latest mountain ranges at any time lying on the margins of continents and not in the interiors. In a very general way, therefore, the continents seem to have grown by the formation of mountain ranges.

While the idea of sea floor spreading seems to describe the motions of sections of the crust and possibly the occurrence of island arcs, the places where mountains form and the growth of continents at the expense of oceanic crust, it leaves unanswered some serious mechanical problems. In order to be able to treat sections of the oceanic areas as well as sections of the continental areas as units in the movements of the surface, it must be supposed that despite the sharp boundary of the Mohorovičić discontinuity the crustal layers are effectively welded to the upper part of the mantle and move with it, forming a layer called the *lithosphere* thought to be about a hundred kilometres thick. The material below the lithosphere to a depth of 200–300 km is called the *asthenosphere* and is supposed to yield readily to long continued stresses. Independent evidence for the distinction is indirect and is not universally accepted, but it is supposed to be consistent with the downwards increase of temperature in the Earth as a consequence of which the mechanical strength of the materials must decrease with depth as indicated in chapter 5, perhaps with partial melting. On the other hand, the existence of isostatic balance seems to indicate that the crust behaves as a rigid plate and the mantle as a material that can flow—on a small scale the regional bending of the oceanic crust around isolated islands and the recovery of the continental crust from the removal of the load of an ice cap seem to require such a model, while on the large scale the relatively local nature of the compensation appears to imply a weak mantle immediately below the crust. Furthermore, the structure of island arcs and the course of formation of a mountain range suggest that, when the crust does yield under compressional stress, the mantle does not bend with it but flows out from underneath it. It seems then that we have two models of the upper layers of the Earth; on the one hand the plate model of the movements of the crust and some seismological results suggest that the crust and the upper part of the mantle move together as a rigid unit, while on the other hand the inference drawn from isostatic structures and the formation of mountains is that the mantle flows more readily than the crust. A related problem is the way in which the world-wide isostatic balance is brought about. The essential point is that the crust and mantle below the oceans are in close balance with the crust and mantle of the continents and evidently, if oceanic crust is being created at the median ridges of the oceans and is being converted into continental crust in mountain ranges, then both processes must in some way control the properties of the crust so formed that isostatic balance is maintained. If the mantle yields to the load of the crust upon it during the formation of mountains, it is easy to envisage that, as the crust thickens under compression, so it is depressed more and more into the mantle to maintain isostatic balance. Thus it would seem that it is the load of the oceans and the crust below them that is given and that determines the thickness of the crust below a mountain range. What then fixes the thickness of the crust as it is formed at the median ridge? Certainly the ridges are too remote from the continents for the formation of oceanic crust to be controlled in any way by the load of the continents upon the mantle, and it may be that the thickness of the oceanic crust is controlled locally by the chemical processes involved.

10.4 Conclusion

A consistent picture of movements at the surface of the Earth appears to be emerging but it is far from clear what forces are entailed. Is the compressional stress restricted to the neighbourhood of island arcs or is the crust as a whole in compression? Do the crust and the upper mantle move as a mechanical unit or is the Mohorovičić discontinuity a boundary between mechanical properties as well as between materials of different composition? While these questions are unresolved and while we have little idea of the balance of work and energy involved in the formation of oceanic crust and the construction of mountains, it cannot be expected that we can have any very realistic idea of how the movements are driven. Here no more will be done than just indicate some of the questions that have been discussed.

What, to begin with, are the forces that drive the lithosphere or the crust as the case may be? The first suggestion, following earlier ideas before sea-floor spreading was thought of, was that the upper mantle is convecting and that the convective motions of the mantle carried the crust or lithosphere upon them. It is very difficult to test the suggestion for we have only very rudimentary ideas of how convection might occur in the mantle, of what the magnitudes of the forces might be and whether the geometric patterns would agree with those of the mid-oceanic ridges. If convection carries the crust, then the stress in the neighbourhood of the mid-oceanic ridges would be tensional while in the neighbourhood of island arcs it would be compressive. The second suggestion is that the formation of crust at the mid-oceanic ridges forces the crust or lithosphere apart and drives it against other plates at the island arcs; in that case, the crust would everywhere be in compression. The third suggestion, conversely, is that crust is in some way sucked down into the mantle at the island arcs, so that the crust is everywhere in tension.

The second group of problems concerns the source of the energy that drives the movements. Because heat is seen to flow out of the Earth it is natural to suppose that the Earth behaves as some sort of heat engine and that heat energy is in some way, such as convection, converted into the mechanical work done in moving the crust. This view is not universally held and the alternative is that chemical energy is released within the mantle, again perhaps being converted into mechanical work through convection or perhaps more directly through processes at the mid-oceanic ridges or the island arcs. In a book on geophysics it can only be said that at the present time we have neither the theoretical background nor the quantitative knowledge of the energies and forces involved that would enable a discussion to be taken further.

Finally, it must be mentioned that the general scheme of continental drift and sea-floor spreading as set out in this chapter and chapter 8 are denied by some; the evidence has been discussed especially by Beloussov (1969).

Examples for chapter 10

10.1 An observatory in lat. 40°N long. 100°W makes observations by laser of the distance to a reflector on the Moon. Calculate the daily variation of that distance. If the distance can be measured to 1 m, estimate the accuracy with which each of the three coordinates of the observatory could be determined. What influence do the position of the reflector and the orbital motion of the Moon have?

10.2 Show that, if two tectonic plates are separating, then the rate of separation at any point is proportional to $\sin\psi$, where ψ is the great circle distance between the point and the pole of relative rotation.

Two tectonic plates are separating by rotation about a pole at 60°N, 70°E. If the rate of separation at 40°N 60°E is 2 cm y^{-1} calculate that at 20°N 80°E.

Further reading for chapter 10

Beloussov, V. V., 1969, 'Interrelations between the Earth's crust and upper mantle', in *The Earth's Crust and Upper Mantle, Geophysics Monograph 13* (Ed. P. J. Hart) (Washington, D.C.: American Geophysical Union).

Bullard, E. C., 1968, 'Reversals of the Earth's magnetic field', *Phil. Trans. Roy. Soc., Ser. A*, **263**, 481–524.

Froome, K. D., and Essen, L., 1969, *The Velocity of Light and Radio Waves* (London, New York: Academic Press), viii + 157 pp.

Hart, P. J. (Ed.), 1969, *The Earth's Crust and Upper Mantle, Geophysics Monograph 13* (Washington, D.C.: American Geophysical Union).

Oliver, J., Sykes, L. R., and Isacks, B., 1969, 'Seismology and the new global tectonics', *Tectonophysics*, **7**, 527–41.

Vine, F., and Matthews, D. H., 1963, 'Magnetic anomalies over oceanic ridges', *Nature*, **199**, 947–9.

Wilson, J. T., 1965, 'A new class of faults and their bearing on continental drift', *Nature*, **207**, 343–7.

CHAPTER ELEVEN

The Earth among the planets

'tanti ch'i'vidi delle cose belle che porta'l ciel.'
Dante, *Inferno* XXIV

11.1 Introduction

From place to place in preceding chapters, properties of the Earth have been compared with the corresponding properties of planets, and now the time has come to set out such comparisons systematically in order to see how similar the Earth is to other planets in its composition and in its response to various forces, for by so doing it may be possible to understand the Earth better. There are in particular four questions that we may understand better if we compare the Earth with the planets—how the Earth and the planets originated, what causes the magnetic field of the Earth, the temperature and sources of heat within the Earth and the support of loads upon the surface of the Earth.

Much the most attention will be paid to the dynamical behaviour of the Moon and the planets, for dynamical data are almost the only exact material we have for their study. We have come to understand the Moon and planets much better in recent years because of observations of artificial satellites and space probes and through observations with radar. The same is even more true of the little we know of the electrical, thermal and magnetic properties of the Moon and the planets, for here our entire knowledge comes from space probes.

11.2 Mechanical and dynamical properties of the Moon and the planets

Here we consider the size, the mass, the speed of rotation, the gravity field and the moments of inertia. From them the density may be calculated and some idea of the central condensation may be obtained.

Prior to the launching of artificial satellites and space probes, the size of the Earth had been found from the measurements of the lengths of arcs of known angular extent upon the surface, but the results were none too representative of the overall dimensions of the Earth because such measurements can be made only upon the land and the land occupies no more than three-quarters of the area of the Earth. Modern estimates of the size of the Earth depend on three methods that employ radar measurements and satellites. In the first place, the geocentric coordinates of a point on the Earth measured from the centre of mass may be found from the daily variations in the distance of that point from a space probe or satellite, as measured by radar for example, or they may be found from the daily variations of the speed of the satellite or space probe relative to the ground site, as found from the change in frequency of a radio transmission from the space probe or satellite according to the Doppler effect. The principle is indicated in figure 11.1. The daily variation in distance or velocity depends on the perpendicular distance of the ground site from the axis of rotation of the Earth, and therefore on the latitude of the site, while the phase of the variation depends on the longitude of the site; it is expected that even more accurate observations will be made by timing pulses of laser light returned from the cube corner reflectors placed upon the Moon by *Apollo* astronauts (chapters 4 and 10). Geometrical measurements give the coordinates of only a few points on the Earth, but those of many others may be found from the shape of the geoid (chapter 2) and so it is possible to obtain the mean radii and the volume of the Earth. Some care is needed in thinking about the units in which the size of the Earth is measured. The dimensions are always expressed in kilometres, but the actual measured quantities are the times of travel of rays of light or of radio

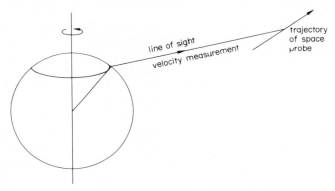

Figure 11.1 Principle of determination of geocentric coordinates from observations of space probes

waves, and the distances in kilometres are calculated with the use of a value of the velocity of light derived from the best laboratory measurements (see Essen and Froome, 1969).

The other modern way of estimating the size of the Earth depends on the relation between the size of a satellite orbit and the period of the satellite in it. Consider, in particular, the motion of the Moon in her orbit. Let the semi-major axis be a and let the mean angular velocity be n. Then

$$n^2 a^3 = GM$$

where M is the mass of the Earth.

n is easily found from long continued observations, while a has been measured with radar.

GM is related to the size of the Earth through the value of gravity upon the surface:

$$g = GM/r^2$$

where r is the mean radius of the Earth.

Eliminating GM,

$$r^2 = n^2 a^3/g$$

In applying this result to find the size of the Earth, careful attention must be given to the meaning of the average values of g and of r, and the effect of the mass of the Moon and of the attraction of the Sun upon the orbit of the Moon have to be taken into account. Nonetheless, the result gives a value of the radius of the Earth that is accurate to a few parts in a million.

The size of the Earth is in fact known far more accurately than is needed in the study of the internal constitution, but the case is different with the Moon and the planets. Prior to the launching of space probes and artificial satellites, the size of any planet was found by multiplying the distance of the planet from the Earth by the angle that it subtended at the Earth and the situation was made more complex by the fact that all distances of planets from the Earth depended on the angle subtended by the Earth at one of the closest planets. The standard of length was thus the diameter of the Earth and all other distances were intercompared through measurements of angles or through the comparison of the periods of orbits of planets about the Sun and the use of Kepler's law. The complexity of these procedures and their dependence upon angular measurements made with telescopes, often under less than ideal conditions, meant that some estimates of the sizes of planets were most uncertain and prevented any detailed discussion of the constitutions. Radar measurements to the planets have now given not only accurate measurements of the distances of the Moon and some of the planets but have also provided values of the sizes (Dollfus, 1970). Coordinates of particular points may also be found from observations of space vehicles that land on the Moon or on a planet and the dimensions can be obtained from observations from an orbiting satellite. The dimensions of the planets are given in table 11.1.

The masses of the planets are best found from satellites according to Kepler's law:

$$GM = n^2 a^3$$

but not all planets have natural or artificial satellites and then other data must be used. The mass of the Earth can be found from the value of gravity at the surface, from the parameters of the orbit of the Moon, or from the orbits of artificial satellites, the semi-major axes of the orbits being found from radar measurements or from radio Doppler shifts. Because the radar or radio measurements are made from the surface of the Earth, they actually measure some combination of the semi-major axis of the orbit with the radius of the Earth and so it is not possible to find the mass of the Earth independently of the size. Like the size, the mass is known to a few parts in a million. The mass of the Moon can be found by comparison with the mass of the Earth or now directly from the orbit of a satellite around her. The comparison with the mass of the Earth depends on the fact that the Earth like the Moon describes an orbit about their common centre of mass

Table 11.1   The mechanical properties of the planets

	Distance from Sun (10^6 km)	Spin period	Radius (km)	Number of satellites	Mass (kg)	Mean density (kg m^{-3})	$J_2 \times 10^{-3}$	$J_4 \times 10^{-6}$	$1/f$	C/Ma^2
Mercury	58	59 d	2434	—	$3{\cdot}3 \times 10^{23}$	5490	—	—	—	—
Venus	108	244 d	6052	—	$4{\cdot}9 \times 10^{24}$	5246	<0·01	—	—	—
Earth	150	1 d	6378	1	6×10^{24}	5517	1·082	1·65	298·25	0·3308
Moon		1 month	1738	5 artificial	$7{\cdot}4 \times 10^{22}$	3342	0·2	—	—	0·40
Mars	228	1 d	3395	2	$6{\cdot}4 \times 10^{23}$	3937	1·97	—	191	0·376
Jupiter	778	10 h	70866	12	$1{\cdot}9 \times 10^{27}$	1363	14·7	−670	15·1	0·264
Saturn	1430	10 h	60016	9	$5{\cdot}7 \times 10^{26}$	696	16·7	−1030	10·2	0·207
Uranus	2870	10 h	25402	5	$8{\cdot}9 \times 10^{25}$	1335	5·0	—	(18)	(0·26)
Neptune	4500	15 h	22300	2	1×10^{26}	1570	—	—	46	0·26
Pluto	5900	6·4 d	3200*	—	$6{\cdot}6 \times 10^{23}$	4800†	—	—	—	—

* Or 2750.
† Or 7900.

and, since the mass of the Earth is about 81 times that of the Moon, the radius of the orbit of the Earth is about 4000 km. Prior to the launching of space probes, the radius of the Earth's orbit was found from the monthly variation of the direction of the Sun among the stars. The stars being at very great distances show no change of apparent direction as the Earth goes round in its little orbit, but the Sun at a distance of $1·5 \times 10^8$ km changes its direction by $2·7 \times 10^{-5}$ rad or some $6''$ and this small change could be measured. Recently, the velocity of the Earth in its orbit has been measured with much greater accuracy. Suppose that a space probe is moving freely away from the Earth. The relative velocity of the Earth and the probe will vary in the course of a month by $2\pi \times 4000$ km per month, or about 10 m s^{-1} and this small change can be measured through the Doppler shift of radio transmissions from the probe. The mass of the Moon so found is, as the following figures show, in very good agreement with that found from an artificial lunar satellite:

mass of Earth/mass of Moon, from orbital velocity of Earth: 81·3008

from orbit of lunar satellite: 81·3022

The size of the lunar satellite orbit is not found, instead the velocity of the satellite in its orbit about the Moon is found from the Doppler shift of radio transmissions. If a is the semi-major axis of the orbit of the satellite, the maximum velocity along the line of sight to the Earth is na; calling this velocity v, Kepler's equation becomes

$$GM = v^3/n$$

Of the planets, Mars has natural satellites as have Jupiter, Saturn, Uranus and Neptune, and from them quite good values of the masses are derived, although the semi-major axes of the orbits have to be found from telescopic measurements of the angular distance of the satellite from the planet. Venus and Mercury have no satellites and in the past their masses have been derived from the quite involved calculations of the effects that their attractions have upon the orbits of other planets about the Sun. The mass of Mercury (and of Pluto) must still be found in this way but the mass of Venus has recently been found from the change of velocity of a space probe travelling in her neighbourhood.

The masses of the planets and of the Moon are given in table 11.1.

The values of the coefficients J_2 in the spherical harmonic expansion of the gravity field of a planet are also best found from the changes in the orbits of satellites, just as for the Earth. However, only four planets have natural satellites that enable a value to be obtained, namely Mars, which has two natural satellites, Jupiter which has 12, Saturn which has 9, and Neptune with 2.* Naturally the changes in the orbits cannot be found so well at a great distance by telescopic observations as they can for satellites about the Earth, and so far only J_2 has been found for any planet, apart from J_4 for Jupiter and Saturn. The value of J_2 for

---

* Uranus also has satellites, but his equatorial plane is perpendicular to the ecliptic, making it difficult to determine J_2.

Venus has also been investigated with the aid of a space probe in the neighbour-hood and has been shown to be very small. The values of J_2 and J_4 for the planets for which they are known are included in table 11.1.

The gravity field of the Moon is complex and cannot be sufficiently described by J_2 alone. It does have some effect on the orbit of the Moon about the Earth and estimates of the coefficients J_2 and C_{22} were made from very small changes in the longitude of the node and perigee of the orbit, but the results for various reasons now seem unreliable. Artificial satellites about the Moon have given far more detailed and satisfactory data. Orbits can be determined in some detail from the Doppler shifts of radio transmissions and, as with satellites about the Earth, can be interpreted in terms of the coefficients of the expansion of the gravitational potential in spherical harmonics. Two features dominate the field of the Moon: first, the three principal moments of inertia of the Moon are unequal, unlike the Earth for which the equatorial ones are almost identical. In consequence the co-efficients J_2 and C_{22} are comparable whereas for the Earth C_{22} is far less than J_2. Secondly, on the Earth, the higher coefficients in the expansion die away much more rapidly with order than they do on the Moon, as is shown in figure 11.2. Maps of the gravity field of the Moon were shown in figures 2.20 and 2.21, and the value of J_2 is given in table 11.1.

The gravity field of a planet by itself tells us something about how the density within the planet varies with angle, but it tells us nothing about how it varies with radius, for which we have to combine the gravity coefficient J_2 with the preces-sional constant H. The Moon, having three distinct moments of inertia, cannot be characterized by a single precessional constant and three ratios α, β and γ are defined:

$$\alpha = (C - B)/A$$

$$\beta = (C - A)/B$$

$$\gamma = (B - A)/C$$

The oscillations of the Moon (librations) under the varying attractions of the Earth and the Sun are controlled by the ratios β and γ as has been indicated in chapter 4, and the ratios may be obtained from an analysis of the librations. The values are

$$\alpha = 3 \cdot 93 \times 10^{-4}$$

$$\beta = 6 \cdot 25$$

$$\gamma = 2 \cdot 32$$

The moment of inertia ratios A/Ma^2, B/Ma^2 and C/Ma^2 follow from the values of J_2, C_{22} and α, β and γ, and are

$$\left.\begin{array}{c} A/Ma^2 \\ B/Ma^2 \\ C/Ma^2 \end{array}\right\} = 0\cdot403 \pm 0\cdot003$$

The expected values for an ellipsoid of constant density differ slightly among each other and from the value of 0·4 for a sphere of uniform density, but the differences are too small for it to be said that the density is other than uniform.

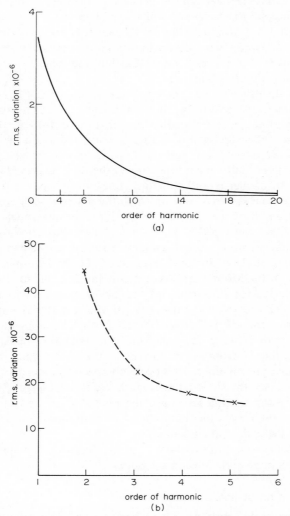

Figure 11.2 Dependence on order of coefficients of harmonic terms in gravitational potentials of (a) Earth and (b) Moon

The foregoing results do need, however, to be treated with some caution. The values of J_2 and C_{22} seem reasonably well established from artificial satellites about the Moon, although the fact that a very large number of harmonic coefficients has been estimated from only five orbits means that the estimates of the different coefficients are related to each other. The values for β and γ, from which α may be deduced by the relation

$$\alpha - \beta + \gamma = \alpha\beta\gamma$$

are not so well established, for they depend on telescopic observations of the position of a small crater (Mösting A) relative to the limb of the Moon, observations which are subject to many difficulties. The values for the moment of inertia ratios just given, although they are the best that can be derived at present, may well change when, for example, the librations have been determined from the laser measurements of distances from the Earth to the cube corner reflectors left upon the Moon by the *Apollo* astronauts.

Precession of the other planets cannot so far be observed. Mars must precess under the attraction of the Sun but the movement is too small to be observed from the Earth although it is possible that in time it will be observed from artificial satellites. (The precession of Jupiter and Saturn can be detected telescopically.) However, we know that the Earth is in a very nearly hydrostatic state, so that the polar flattening is very close to that to be expected for a planet of the same moment of inertia and with the same rate of spin and in a strictly hydrostatic state, that is to say, one in which the material supports no shearing stresses. The strength of the material of the Earth is very much less than the pressure throughout most of the Earth and, if the same condition applies in another planet, then it would seem reasonable to calculate the moment of inertia from the flattening on the hydrostatic assumption. The hydrostatic assumption is not a good one for the Moon, but it is evidently much better for Jupiter, Saturn and Neptune than it is for the Earth, since they are much larger than the Earth and the pressures much greater, while the strength of any material is unlikely to be very different from that of the material of the Earth. The moments of inertia of Jupiter, Saturn and Neptune can therefore be confidently calculated from the polar flattening, that is, from the coefficient J_2, the spin angular velocity being known. The hydrostatic assumption is probably less satisfactory for Mars than it is for the Earth, since the shear stresses are likely to be a larger fraction of the pressure, the pressures being less since the planet is smaller than the Earth. However, in the absence of other data, the moment of inertia ratio for Mars also is calculated on the hydrostatic assumption.

The moment of inertia ratios for Mars, Jupiter, Saturn and Neptune are given in table 11.1. It is impossible to estimate values for Mercury and Venus.

11.3 The zones of the planets

The main features of the structure of the Earth as set out in chapter 5 are that the Earth is divided into two major zones, the core and the mantle, with very different chemical compositions, and that the material of the mantle undergoes a change to a more compact form at a depth of about 400 km, the corresponding pressure being 2×10^{10} N m^{-2} and, secondly, that throughout the Earth the bulk modulus K is very nearly a linear function of pressure. In detail the picture is too simple and it may well be that there are changes of composition as well as of phase within the mantle, but the simple scheme of a phase change in the mantle combined with the behaviour of the bulk modulus is so successful and of such generality that clearly it should be applied also to the planets for which the only observed data we have are the mechanical quantities described in the previous section. The models that have been constructed for the Earth have been obtained from seismic as well as dynamical data and for the planets as for the Earth dynamical data by themselves are insufficient to enable models to be constructed. By making use of the rule for the variation of bulk modulus with pressure and of the idea of the phase change with pressure to derive models for the planets, we are saying in effect that the planets are all generally similar to the Earth and that if we were able to make seismic observations upon them we would get much the same results as we do upon the Earth.

It is possible that, if seismic observations were attempted upon the planets that we should find that they did not give the same results as upon the Earth, for we already know that that happens on the Moon. The seismic signals detected by seismometers placed upon the Moon by the *Apollo* astronauts are long trains of waves quite different from the discrete pulses seen from earthquakes on the Earth and they last for so long that they would obscure any discrete pulses that could give information about a zonal structure of the Moon. Mars, which has surface features like the Moon, might conceivably give seismic signals more like those on the Moon than those on the Earth.

Let us, however, assume that the linear variation of bulk modulus with pressure and the change of phase under pressure are such essential properties of the sort of material that planets are made of that we can suppose that they apply to the planets as to the Earth. The following equations then apply within a zone of a planet in which there is no change of phase:

$$\text{dependence of density on pressure: } \mathrm{d}\rho/\mathrm{d}p = \rho/K \qquad (11.1)$$

$$\text{variation of pressure with radius: } \mathrm{d}p/\mathrm{d}r = -GM_r\rho/r^2 \qquad (11.2)$$

$$\text{mass of a spherical shell: } \mathrm{d}M_r = 4\pi\rho r^2\,\mathrm{d}r \qquad (11.3)$$

$$\text{dependence of bulk modulus on pressure: } K = a + bp \qquad (11.4)$$

ρ is the density at radius r and M_r is the mass within radius r. Combine equation (11.1) with equation (11.4)

$$\frac{\mathrm{d}\rho}{\rho} = \frac{\mathrm{d}p}{a + bp} \qquad (11.5)$$

The solution of equation (11.5) is

$$\ln(a + bp) = b\ln\rho + \text{constant} \qquad (11.6)$$

that is

$$\rho = \rho_0(a + bp)^{1/b} \tag{11.7}$$

or

$$K = A\rho^b \tag{11.8}$$

Then

$$\frac{d\rho}{dr} = \frac{d\rho}{dp}\frac{dp}{dr} = -\frac{GM_r\rho^2}{A\rho^b r^2}$$

$$= -\frac{G}{A}\frac{M_r}{r^2}\rho^{2-b}$$

or

$$M_r = -\frac{A}{G}r^2\,\rho^{b-2}\frac{d\rho}{dr} \tag{11.9}$$

But

$$\frac{dM_r}{dr} = 4\pi\rho r^2$$

and so

$$\frac{d}{dr}\left(\frac{A}{G}r^2\,\rho^{b-2}\frac{d\rho}{dr}\right) = -4\pi\rho r^2$$

or

$$\frac{d}{dr}\left(r^2\,\rho^{b-2}\frac{d\rho}{dr}\right) = -Br^2\,\rho \tag{11.10}$$

where

$$B = 4\pi G/A$$

Equation (11.10) is an example of a type of differential equation known as Emden's equation, an equation that occurs in the theory of the internal structure of stars. It must in general be solved numerically and, in particular, numerical solutions are needed for the planets for which, by analogy with the Earth, b should be about 3·5. Suppose in the first place that no phase change or change of composition take place within a planet. Then, if a value of the density is assumed at the outer surface, Emden's equation may be integrated numerically to give the density as a function of radius throughout the planet, and it is therefore possible to calculate the radius and the moment of inertia for a planet of given mass. The pressure in the Moon nowhere exceeds that at 400 km in the Earth and it is therefore reasonable to suppose that no phase change occurs within the Moon and to see how well the model of uniform composition fits the Moon. In fact, if the density at the surface of the Moon is taken to be the same as that assumed for the top of the Earth's mantle, then the calculated properties of the Moon agree very well with the observed properties as shown in the following list.

	Observed	Calculated
Radius	1738 km	1737 km
Moment of inertia ratio	0·400	0·398

The dynamical properties are thus consistent with the idea that the Moon is made of material with properties similar to those of the upper mantle of the Earth.

The pressure in Mars may be expected at some depth to attain that at 400 km in the Earth and so it is necessary to consider how to integrate Emden's equation if a change of phase or composition with consequent increase of density is supposed to occur. In that case the integration gives the density as before down to the depth at which the phase change takes place. The integration must then be started again for the inner zone. The density is now not given but must be calculated from the final value of the first integration increased by the supposed change of density on passing to the denser zone, a value taken from the results for the Earth. Integration then proceeds as before to the centre of the planet. Again, the calculations give the density throughout the model, from which the radius and moment of inertia for a given mass ($6·4 \times 10^{23}$ kg) may be calculated, but now it is possible to choose not only the overall radius of the planet but also the pressure at which the change of phase is supposed to occur. It is not necessarily correct to choose the same pressure as in the Earth because the transition pressure of a phase change is affected to some extent by the temperature and that may well not be the same in Mars as in the Earth, while any change of composition will be independent of pressure. Thus results for a model of Mars cannot be as certain as those for a model of the Moon and, because the pressure at which the phase transition may occur is variable, the radius and moment of inertia ratio are not both determined as for the Moon. If the mass is fixed, there will be a set of pairs of radii and moment of inertia ratios corresponding to different pressures for the phase transition; however, the least value of C/Ma^2 for any such pressure is found to be 0·380, whereas the value based on the observed value of J_2 is 0·376. Mars is therefore more concentrated at the centre than any such model would entail and so may have a small central heavy core.

Because the moments of inertia of Venus and of Mercury cannot be found, it is not possible to check solutions of Emden's equation in the same way as for the Moon and for Mars. It does, however, seem clear that the mass of Venus cannot be reproduced unless she is supposed to have a small iron core similar to that of the Earth. Mercury, not much larger than the Moon, is also too dense to be composed of material similar to the mantle of the Earth but, in the absence of any information about the central condensation, it is not possible to say anything useful about the composition.

The major planets, Saturn and Jupiter, have densities very much less than those of the Earth and its near neighbours, and the compositions are therefore quite different. Very simple calculations show that, at the pressures that must be attained in the major planets, the densities of any materials except those composed predominantly of hydrogen would exceed the observed densities. It would of course be possible to attempt to model the major planets with Emden equations, but the small values of the moment of inertia ratios show that the planets are very strongly condensed indeed towards the centre, suggesting that there exist cores with changes of phase or of composition. Nothing that we know about the Earth would be of any value as a guide to the changes of density to be assumed at the boundaries of such cores, and so in studying the major planets it is necessary to attempt to calculate the properties of the material theoretically from first principles. Then it is also possible to calculate the change of density with pressure throughout the planet without recourse to the empirical Emden's equation.

Calculations of the properties of Jupiter and Saturn are based on the supposition that the planets consist of hydrogen, with perhaps 10% of helium added, as in the typical composition of stars. Quantum-mechanical calculations indicate that at a pressure of about 10^{12} N m^{-2} hydrogen changes from a molecular gas to a metal in which free electrons form a gas within a lattice of protons. Hydrogen in the metallic state is the simplest of all metals and when its properties are calculated by quantum mechanics they should be fairly reliable.

In principle the calculation would involve solving Dirac's equation for N electrons in a lattice of N protons. That is at present beyond the scope of practical calculation but by a series of approximations it is likely that a quite reliable result has been obtained.

The first step is to solve Schrödinger's equation with a central potential $(-e^2/r)$ just as for the hydrogen atom, only instead of requiring that the wave function, ψ, should vanish at infinity it is confined to a sphere of radius r, the radius corresponding to the atomic volume, by requiring that $\partial\psi/\partial r$ shall vanish on that sphere. This is similar to the procedure in the Thomas–Fermi–Dirac theory (chapter 5).

Secondly, the energy of the electron gas proportional to $1/r^2$ is included.

Thirdly, corresponding to the fact that electrons are indistinguishable there is an exchange energy of form $a/r - b$ and because electrons move collectively in the lattice (as plasma oscillations at ordinary temperatures or supercurrents at low temperatures) there is a correlation energy $-\alpha(r + \beta)$. Finally, the zero point energy must be included.

The following figures give an idea of the results of such calculations (which apply to 0 °K).

p (10^{11} N m^{-2})	ρ (kg m^{-3})
0·052	550
0·59	730
2·5	1100
8·8	1800
22·4	2700

The properties of molecular hydrogen cannot be calculated from first principles because molecular hydrogen is a much more complicated system than metallic hydrogen, but an equation of state has been determined by experiment up to 2×10^9 N m^{-2} and has been extrapolated to higher pressure in a more or less empirical way. Thus it has been possible to estimate the range of pressure in which the transition to the metallic form is likely to occur.

The results for one possible model of Jupiter are shown in figure 11.3. If the addition of 10 % of helium is admitted, it is possible to get quite a good match to the mass and moment of inertia of Jupiter, but the models of Saturn are less

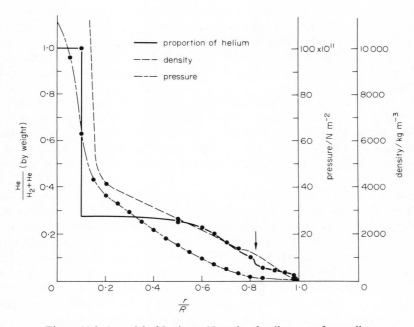

Figure 11.3 A model of Jupiter: r/R, ratio of radius to surface radius

satisfactory. Although smaller, so that the pressures are less than in Jupiter, Saturn is more strongly condensed, as shown by a smaller moment of inertia ratio, and so it is likely that it must contain some material appreciably denser than hydrogen or helium in a central core.

While the simple models of hydrogen planets give a good first idea of the composition and properties of the major planets, there are important features which are unexplored. In particular, in view of the fact that, as will be seen, both planets have magnetic fields, it would be very desirable to know whether the metallic cores are solid or liquid; there is indeed the possibility that they are so large that they might be superconducting.

11.4 Irregularities

We know well that a simple model with zones of uniform composition and properties that depend only on the radius is an inadequate description of the Earth and we may suppose that, had we more information about the planets, we should find that it was inadequate for them also. The model for the Earth breaks down in two ways—seismic evidence shows that there are relatively slow changes of composition within the major zones, especially within the upper mantle, and various observations show us (see chapter 5) that the upper parts of the Earth are not in a hydrostatic state but that they support shearing stresses and that there are corresponding variations of density and other properties of the upper mantle with angular direction, variations that would not be there if the Earth were in a hydrostatic state. Seismic evidence is not available for the planets but observations of the surface, together with a rather detailed knowledge of the gravity field of the Moon, give us some idea of corresponding features of the planets.

The very fact that the solid surface of a planet is irregular shows that the material can support shearing stresses, and direct observations show that the surface of the Moon has irregularities of up to 2 km, while the irregularities on Mars may amount to 15 km. The corresponding strength of the material of the Moon and of Mars must be at least as great as that of the mantle of the Earth, say 10^7 N m^{-2}.

The gravity field of the Moon has already been seen to be relatively more irregular than that of the Earth and figure 11.2 shows that the irregularities at the shorter wavelengths do not die away as do those on the Earth, behaviour which suggests that the variations of density that are responsible for the irregularities of gravity lie closer to the surface of the Moon than do the corresponding variations in the Earth. We have seen (chapter 5) that the variations of density in the Earth are not, on the whole, associated with the continents and oceans but probably lie in the upper mantle where they may be supported by the strength of the material or by movements in the mantle. If the strength of the material supports the loads then the strength must be at least 10^7 N m^{-2}, about the same as is required to support the amount by which the equatorial bulge exceeds the value corresponding to the spin of the Earth in the hydrostatic state. If the loads on the Moon are near the surface, then the strength required in the material of the Moon is again about 10^7 N m^{-2} and again that is much what is needed to support the equatorial bulge of the Moon. The most striking loads at the surface of the Moon are the mascons (chapter 2) associated with the larger maria.

The data that we have are thus consistent with the idea that the materials of the Moon and the Earth have similar strengths, at least near the surface, and that the variations of density with angular position are comparable. We do not at present have sufficient data to say anything quantitative about the other planets but the heights of topography that have been measured on Mars suggest that the loads are greater than on the Moon or on the Earth.

11.5 Magnetic, electrical and thermal properties

Our knowledge of the magnetic, electrical and thermal properties of the Moon and, so far as it goes, of the other terrestrial planets, comes from magnetometers carried in spacecraft to the vicinity of those bodies. In fact, those measurements show that there are no detectable permanent dipole fields in the neighbourhoods of any of those bodies and that the Earth is the only one among its neighbours that has a permanent field. Jupiter, on the other hand, does possess a dipole field. It emits radio waves which originate in electrical discharges in its atmospheres and from the properties of the radiation it has proved possible to infer that the planets have general fields of the order of 5×10^{-3} T.

The lack of general dipole fields around the Moon, Mars and Venus can be understood if the dynamo theory is accepted for the origin of planetary fields, for that theory requires that, to possess a general field, a planet should have a fluid core that is electrically conducting and it suggests that the planet should be spinning fairly rapidly. The Moon and Mars are unlikely to have metallic cores and Venus, although it may well have a metallic core, is spinning very slowly. We do not know if Mercury has a general field but it also is spinning very slowly and so it is unlikely that it does have a field. Jupiter and Saturn spin faster than any other planet and, according to the ideas set out above, they have cores of metallic hydrogen; however, we do not know whether theory predicts that the cores would be solid or liquid, and there is the possibility that they might be superconducting, with a field frozen in.

Magnetometers taken on to the Moon by the *Apollo* astronauts have shown that there are small local fields that may be due to the permanent magnetization of some of the surface material. The interpretation is as yet obscure—it has, for example, been suggested that the Moon originated as part of a larger body that did possess a field.

When magnetometers are carried in space vehicles in interplanetary space, they detect small fields caused by and frozen into the streams of charged particles, the solar wind, that flow out from the Sun. The fields are not constant but vary from time to time and from place to place as a result of variations in the flow of particles from the Sun and of variations in the solar field at the place whence the solar wind came. It has been found in one instance that, when such an irregularity in the field passed by the Moon, it was slightly distorted and it has been inferred that the distortion was caused by electromagnetic induction in the Moon. Thus it was possible to estimate the electrical conductivity of the Moon, a value of less than 10^{-4} mho m^{-1} being found. If it is supposed that the Moon is composed of the same type of semiconducting silicate material as the upper mantle of the Earth, the fact that the electrical conductivity of the Moon appears to be very much less than that of the upper mantle implies that the temperature of the Moon is much lower than that of the Earth and probably nowhere exceeds 1000 °K. It must be emphasized that we are here in the realms of speculation but it is very much to be hoped that it will be possible to make further observations of electromagnetic induction in the Moon and the planets so that the electrical conductivity may be more certainly established and some definite idea obtained of the internal temperature.

It does not seem likely that it will be possible to make direct thermal measurements on the Moon, far less on the planets, for some time to come, although tentative experiments have been made on the Moon by the *Apollo* astronauts and by the Russian unmanned lunar vehicles, and the only data we can expect will come from estimates of the electrical conductivity. It is evident from what has been said about the support of irregular loads on the Earth by the strength of the material on the one hand or by convection on the other, from what has been said about the generation of the general magnetic field of the Earth and from what has been said about the Earth as a heat engine, that some knowledge of the temperature within a planet is essential to understanding its internal state and how it has come to its present condition.

11.6 Conclusion

The launching of spacecraft has made it possible for the first time to extend to the planets the methods of physical study that have been applied to the Earth. Already, we have quite good knowledge of the dynamics of many of the planets and can see which are strongly condensed and have cores and those which show little or no central condensation.

Thus it begins to be possible to see how to construct models of the terrestrial planets on the basis of properties of materials with which we are familiar in the Earth. At first we are restricted to studying the planets as cold bodies, in the sense of chapter 5—bodies in which the properties of matter are determined by chemical composition and hydrostatic pressure. But spacecraft observations have already shown the Moon and the nearer planets to have quite large surface irregularities which must both be supported by departures from the hydrostatic state and which were brought into being most likely by thermal forces. If movements are still going on in a planet, then there is some likelihood of learning more about its detailed internal structure from records of seismometers placed upon it by spacecraft and, if there is still movement, then it is the more important to try to estimate the flow of heat through the planet.

It would be quite wrong to end this chapter, as it would be wrong to end the book, with the impression that we have little more to learn in understanding the Earth and the planets. On the Earth, problems of thermodynamics are still far from understood, while for the inner planets we have reached the stage where we can see that developing analogies with the Earth is likely to be a very fruitful means of study; as for the major planets, again enough has been done to suggest that the ways in which we investigate those planets are on the right lines.

There is thus much to do, and a wide field of study, of intense interest not only for the natural history of the Earth and its companions, but also for the way in which it extends our understanding of physics by presenting to us materials under conditions that we cannot achieve in the laboratory.

Further reading for chapter 11

Cook, A. H., 1972, 'The dynamical properties and internal structures of the Earth, the Moon and the planets', *Proc. Roy. Soc., Ser. A*, **328**, 301–36.

Dollfus, A. (Ed.), 1970, *The Surfaces and Interiors of Planets and Satellites* (London, New York: Academic Press).

Froome, K. D., and Essen, L., 1969, *The Velocity of Light and Radio Waves* (London, New York: Academic Press), viii + 157 pp.

Jeffreys, H., 1970, *The Earth*, 5th edn (Cambridge: Cambridge University Press).

Lyttleton, R. A., 1965, 'On the internal structure of the planet Mars', *Monthly Notices Roy. Astron. Soc.*, **129**, 21–39.

Note added in proof

(1) A possible interpretation of the magnetization of lunar rocks is that the Moon has a small metallic core which sustained dynamo action early in lunar history.

(2) Heat flow measurements have been made by *Apollo* astronauts and give an estimate of 0.03 W m^{-2}, half the rate at which heat flows out of the surface of the Earth.

Appendixes

Some numerical data

Constants

Constant of gravitation	G	6.67×10^{-11} N kg^2 m^{-2}
Speed of light	c	2.997×10^8 m s^{-1}
Boltzmann's constant	k	1.3805×10^{-23} J degK^{-1}
Stefan's constant	σ	5.67×10^{-8} J m^{-2} s^{-1} degK^{-4}
Avogadro's number		6.0226×10^{26} kg^{-1} mole^{-1}

The Earth

Equatorial radius	a	6378.160 km
Polar radius	b	6356.78 km
Polar flattening	f	$1/298.254$
Volume		1.0832×10^{21} m^3
Mass		5.977×10^{24} kg
GM		$3.986\,03 \times 10^{14}$ m^3 s^{-2}
Mean density		5517 kg m^{-3}
Areas: surface		5.1×10^{14} m^2
oceans		3.63×10^{14} m^2
continents		1.48×10^{14} m^2
Dynamical ellipticity:	H	3.2756×10^{-3}
	J_2	1.0827×10^{-3}

Moments of inertia $C/Ma^2 = 0.3306$; $A/Ma^2 = 0.3295$

Gravity: γ_e, equatorial value at sea level $\quad 9.780\,318$ m s^{-2}

Standard gravity at sea level

$$\gamma = \gamma_e(1 + \beta_1 \sin^2 \phi + \beta_2 \sin^2 2\phi)$$

where ϕ is the *geographic latitude*

$$\beta_1 = 5.3024 \times 10^{-3}$$
$$\beta_2 = 5.9 \times 10^{-6}$$

Angular velocity of the Earth	$7.292\,115 \times 10^{-5}$ rad s^{-1}
Length of the solar day	$86\,400$ s
Length of the year	3.1558×10^7 s
Radius of Earth's orbit	1.4960×10^8 km
(astronomical unit, A.U.)	
Radius of Moon's orbit	3.84×10^5 km
Mass of Sun	1.99×10^{30} kg
Mass of Moon	7.35×10^{22} kg
	$= 1/81.303 \times$ mass of Earth

Mass of atmosphere	5.27×10^{18} kg
crust	2.4×10^{22} kg
mantle	4.1×10^{24} kg
oceans	1.24×10^{21} kg
Age of the Earth	4.5×10^{9} y
Greatest depth of oceans	10·550 km
Mean depth of oceans	3·88 km
Greatest height of land	8·84 km
Mean height of land	0·84 km

Earth's core

Radius	3476 km
Volume	1.76×10^{20} m^3
Mass	1.88×10^{24} kg
Mean density	10 720 kg m^{-3}
Moment of inertia, C/Ma^2	0·380

Solution of Laplace's equation in spherical polar coordinates

In spherical polar coordinates (r, θ, ϕ) Laplace's equation, $\nabla^2 V = 0$, reads

$$\frac{\partial^2 V}{\partial r^2} + \frac{2}{r}\frac{\partial V}{\partial r} + \frac{1}{r^2 \sin\theta}\frac{\partial}{\partial\theta}\left(\sin\theta\frac{\partial V}{\partial\theta}\right) + \frac{1}{r^2 \sin^2\theta}\frac{\partial^2 V}{\partial\phi^2} = 0$$

Suppose that V can be expressed as the product of functions that depend respectively on r, θ, ϕ, alone:

$$V = R(r)\,\Theta(\theta)\,\Phi(\phi)$$

If R' denotes $\partial R/\partial r$, R'' denotes $\partial^2 R/\partial r^2$, Θ'' denotes $\partial^2\Theta/\partial\theta^2$, and so on, Laplace's equation reads

$$\Theta\Phi\left(R'' + \frac{2R'}{r}\right) + \frac{R\Phi}{r^2 \sin\theta}\frac{\partial}{\partial\theta}(\sin\theta\,\Theta') + \frac{R\Theta}{r^2 \sin^2\theta}\,\Phi'' = 0$$

Now multiply by $r^2 \sin^2\theta/V$. Then

$$\frac{r^2 \sin^2\theta}{R}\left(R'' + \frac{2R'}{r}\right) + \frac{\sin\theta}{\Theta}\frac{\partial}{\partial\theta}(\sin\theta\,\Theta') + \frac{\Phi''}{\Phi} = 0$$

The first two terms are functions of r and θ but not of ϕ, while the third term is a function of ϕ alone. If the equation is to be satisfied by all values of r, θ, ϕ the last term and the first two must respectively be independent of r, θ, ϕ. Thus

$$\frac{\Phi''}{\Phi} = -m^2, \text{ a constant}$$

or

$$\Phi = e^{\pm im\phi}$$

Now $\Phi(\phi + 2\pi)$ must be identical with $\Phi(\phi)$ and so $e^{\pm im2\pi} = 1$, or m must be a real integer.

The remaining terms of Laplace's equation then read

$$\frac{r^2 \sin^2\theta}{R}\left(R'' + \frac{2R'}{r}\right) + \frac{\sin\theta}{\Theta}\frac{\partial}{\partial\theta}(\sin\theta\,\Theta') - m^2 = 0$$

Divide by $\sin^2 \theta$:

$$\frac{r^2}{R}\left(R'' + \frac{2R'}{r}\right) + \frac{1}{\Theta \sin \theta} \frac{\partial}{\partial \theta}(\sin \theta \, \Theta') - \frac{m^2}{\sin^2 \theta} = 0$$

Again, the first term is a function of r alone, the second and third of θ alone and so for the equation to be satisfied for all r and θ each must be a constant. Hence

$$\frac{1}{\Theta \sin \theta} \frac{\partial}{\partial \theta}(\sin \theta \, \Theta') - \frac{m^2}{\sin^2 \theta} = -C$$

and

$$\frac{r^2}{R}\left(R'' + \frac{2R'}{r}\right) = +C$$

The equation for Θ then reads

$$\frac{1}{\sin \theta} \frac{\mathrm{d}}{\mathrm{d}\theta}(\sin \theta \, \Theta') + \left(C - \frac{m^2}{\sin^2 \theta}\right)\Theta = 0$$

Real solutions on a sphere are obtained if $C = n(n + 1)$.
The equation for R then becomes

$$r^2 R'' + 2rR' - n(n + 1)\, R = 0$$

of which the solutions are

$$R = r^n \text{ or } r^{-n-1}$$

If the sources of a field are inside a sphere of radius a, the field must vanish at infinity and the solution will be

$$R = \left(\frac{a}{r}\right)^{n+1}$$

The gravitational field is of this form.
If the sources are outside a sphere, the field must vanish at the centre of the sphere and the solution will be

$$R = \left(\frac{r}{a}\right)^{n}$$

Part of the geomagnetic field is of this form.

The equation for Θ reads

$$\frac{1}{\sin\theta}\frac{d}{d\theta}(\sin\theta\,\Theta') + \left\{n(n+1) - \frac{m^2}{\sin^2\theta}\right\}\Theta = 0$$

It is *Legendre's* equation and the solutions are *Legendre's polynomials*, $P_n(\cos\theta)$, if $m = 0$, and *Legendre's functions*, $P_n^m(\cos\theta)$, if m is not zero (see Whittaker and Watson, 1940, *Modern Analysis* (Cambridge: Cambridge University Press)).

Some values are

$$P_0: 1$$
$$P_1: \cos\theta$$
$$P_2: \tfrac{1}{2}(3\cos^2\theta - 1)$$
$$P_3: \tfrac{1}{2}(\cos^3\theta - 3\cos\theta)$$

$P_n(\cos\theta)$ is an even function of $\cos\theta$ if n is even and is an odd function if n is odd.

$$P_1^1: \sin\theta$$
$$P_2^1: -3\cos\theta\sin\theta$$
$$P_2^2: 3\sin^2\theta$$

The complete solution of Laplace's equation is thus the sum of terms such as

$$\left(\frac{a}{r}\right)^{n+1} P_n^m(\cos\theta)(\sin m\phi,\ \cos m\phi)$$

and

$$\left(\frac{r}{a}\right)^n P_n^m(\cos\theta)(\sin m\phi,\ \cos m\phi)$$

The combinations

$$P_n^m(\cos\theta)(\sin m\phi,\ \cos m\phi)$$

are known as *surface harmonics*.

If $m = 0$, the harmonic is independent of ϕ and is called a *zonal harmonic*.

If $m = n$, the harmonic has no zeros between $\theta = 0$ and π but vanishes on n great circles between $\phi = 0$ and 2π; it is known as a *sectorial harmonic*.

The general harmonic which vanishes on circles of latitude and on meridians is known as a *tesseral harmonic*.

Zonal harmonics divide the surface of the sphere into zones of latitude, sectorial harmonics divide it into sectors of longitude, whilst tesseral harmonics divide it into 'rectangles' (*tesserae*) by latitude and longitude.

Any arbitrary function given over the surface of a sphere (for example, the value of gravity or a component of the magnetic field) may be expressed as the sum of a set of surface harmonics.

McCullagh's theorem

Let OZ (figure A3.1) be the polar axis of the Earth and let OX be any axis in the equatorial plane. Let P' be any point in the Earth with spherical polar coordinates r', θ', ϕ' and let the density at P' be σ. The moment of inertia about the polar axis is equal to the integral of the product of the mass $\sigma\,dv$ of an element of volume dv multiplied by the square of its distance from the polar axis, that is

$$C = \int dv\,\sigma r'^2 \sin^2\theta' \qquad (A3.1)$$

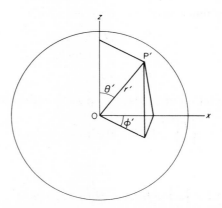

Figure A3.1 Geometry for calculation of moments of inertia of a nearly spherical planet

where the integral is taken throughout the whole volume of the Earth. Since

$$dv = r'^2 \sin\theta'\,dr'\,d\theta'\,d\phi'$$

$$C = \int_0^a dr' \int_0^\pi d\theta' \int_0^{2\pi} d\phi'\,\sigma r'^4 \sin^3\theta'$$

Suppose that, as is very nearly the case, the Earth is symmetrical about the polar axis, so that all directions OX in the equatorial plane are equivalent. The density does not depend on ϕ' and so the integration with respect to ϕ' may be carried out immediately:

$$C = 2\pi \int_0^a dr' \int_{-1}^{+1} \sigma r'^4 \sin^2\theta'\,d(\cos\theta') \qquad (A3.2)$$

The moment of inertia about OX is similarly found by multiplying the element of mass by the square of the perpendicular distance, $r'^2(\cos^2\theta' + \sin^2\theta'\cos^2\phi')$, from OX:

$$A = \int dv\,\sigma r'^2(\cos^2\theta' + \sin^2\theta'\cos^2\phi') \qquad (A3.3)$$

Integration with respect to ϕ' now gives

$$A = 2\pi \int_0^a dr' \int_{-1}^{+1} \sigma r'^4(\cos^2\theta' + \tfrac{1}{2}\sin^2\theta')\,d(\cos\theta') \qquad (A3.4)$$

Hence

$$C - A = -2\pi \int_0^a dr' \int_{-1}^{+1} \sigma r'^4\,\tfrac{1}{2}(3\cos^2\theta' - 1)\,d(\cos\theta')$$

$$= -2\pi \int_0^a dr' \int_{-1}^{+1} \sigma r'^4\,P_2(\cos\theta')\,d(\cos\theta') \qquad (A3.5)$$

because

$$\tfrac{1}{2}(3\cos^2\theta' - 1) = P_2(\cos\theta') \quad \text{(appendix 2)}$$

Now calculate the potential of the Earth at some external point P with co-ordinates r, θ, ϕ (figure A3.2). The potential due to the element of mass at P' will be

$$G\sigma\,dv/D$$

where D is the distance between P and P':

$$D^2 = r^2 + r'^2 - 2rr'\cos\chi$$

χ is the angle POP'.

The potential due to the whole Earth is therefore

$$-G \int \frac{\sigma\,dv}{(r^2 + r'^2 - 2rr'\cos\chi)^{1/2}}$$

$$= -\frac{G}{r} \int \sigma\,dv \left\{ 1 - \frac{r'}{r}\cos\chi + \left(\frac{r'}{r}\right)^2 \tfrac{1}{2}(3\cos^2\chi - 1) + \ldots \right\} \qquad (A3.6)$$

Because the mass of the Earth is just $\sigma\,dv$, the first term in the series is $-GM/r$, the potential of a spherically symmetrical body of the same mass as the Earth.

Since $\cos\chi = \cos\theta\cos\theta' + \sin\theta\sin\theta'\cos(\phi - \phi')$, the second term in the series is

$$\frac{G}{r^2} \int dv\,\sigma r'\{\cos\theta\cos\theta' + \sin\theta\sin\theta'\cos(\phi - \phi')\}$$

Now, if the origin is taken at the centre of mass, $\int dv\,\sigma r'\cos\theta'$ is zero, while

$$\int dv\,\sigma r'\sin\theta'\cos(\phi-\phi')$$

vanishes on account of the symmetry about the polar axis. The second term in the series is therefore zero under these conditions.

The third term is

$$-\frac{G}{r^3}\int dv\,\sigma r'^2 P_2(\cos\chi)$$

Now

$$\cos^2\chi=\cos^2\theta\cos^2\theta'+2\cos\theta\cos\theta'\sin\theta\sin\theta'\cos(\phi-\phi')$$
$$+\sin^2\theta\sin^2\theta'\cos^2(\phi-\phi') \qquad\qquad (A3.7)$$

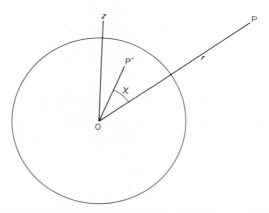

Figure A3.2 Geometry for calculation of potential outside nearly spherical planet

and on substituting this result into the third term and carrying out the integration with respect to ϕ, the term is reduced to

$$-2\pi\frac{G}{r^3}\int_0^a dr'\int_{-1}^{+1} d(\cos\theta')\,\sigma r'^4 P_2(\cos\theta')P_2(\cos\theta) \qquad (A3.8)$$

Now insert the value for $C-A$ from (A3.5) in (A3.8); it follows that the potential is

$$-\frac{GM}{r}+\frac{G(C-A)}{r^3}P_2(\cos\theta) \qquad\qquad (A3.9)$$

showing that J_2 in the general solution of Laplace's equation (2.9) is $(C-A)/Ma^2$.

This is McCullagh's theorem and is a key result in the study of the physics of the Earth.

APPENDIX FOUR

An elementary theory of the precession of an artificial satellite caused by the equatorial bulge

Consider an artificial satellite in a circular orbit about the centre of the Earth. Gauss showed that, if we are only concerned with the changes in the motion of the satellite in the course of many revolutions, then the actual satellite may be replaced by a ring of radius equal to the radius of the orbit, spinning about the centre at the same angular velocity as the satellite in its orbit.

The geometry was shown in figure 2.17. E is the pole of the equator of the Earth and P is the pole of the orbit of the satellite. O is the centre of the Earth and EOP is the angle of inclination of the orbit to the equator (θ). The planes of the equator and the orbit intersect in the line of nodes, NN′, the direction of which makes the angle Ω with the direction, fixed in space, of the first point of Aries (Υ). OX is any axis in the plane of the orbit and rotating at the angular velocity, n, of the satellite in its orbit. OX makes the angle ϕ with ON. θ, Ω, ϕ are Eulerian angles for the specification of the position of the ring equivalent to the satellite.

The direction OP is a principal axis of inertia of the ring and the moment of inertia about it will be denoted by C. C is in fact equal to mr^2 where m is the mass of the satellite and r is the radius of the orbit. The moments of inertia about axes in the plane of the orbit are all the same. They will be denoted by A and are equal to $\frac{1}{2}mr^2$. Let the axes OP, OX be labelled respectively 3 and 1 and let the third mutually perpendicular axis be labelled 2. The angular velocities about the axes are

$$\left. \begin{aligned} \Omega_1 &= \dot{\Omega} \sin\theta \sin\phi + \dot{\theta} \cos\phi \\ \Omega_2 &= \dot{\Omega} \sin\theta \cos\phi - \dot{\theta} \sin\phi \\ \Omega_3 &= \dot{\Omega} \cos\theta + \dot{\phi} \end{aligned} \right\} \qquad \text{(A4.1)}$$

(Dots denote differentiation with respect to time.) If the moments of inertia about the axes 1, 2 and 3 are respectively A, B and C, the equations of motion are

$$\left. \begin{aligned} A\dot{\Omega}_1 + (C - B)\Omega_2 \Omega_3 &= K_1 \\ B\dot{\Omega}_2 + (A - C)\Omega_1 \Omega_3 &= K_2 \\ C\dot{\Omega}_3 + (B - A)\Omega_1 \Omega_2 &= K_3 \end{aligned} \right\} \qquad \text{(A4.2)}$$

where K_1, K_2, K_3 are the couples about the respective axes.

Now for the satellite ring, A equals B, there is no preferred axis in the equator, and ϕ may be taken as instantaneously zero. Then

$$\left.\begin{array}{ll}\Omega_1 = \dot{\theta}, & \dot{\Omega}_1 = \dot{\Omega}\dot{\phi}\sin\theta + \ddot{\theta} \\[4pt] \Omega_2 = \dot{\Omega}\sin\theta, & \dot{\Omega}_2 = \ddot{\Omega}\cos\theta + \dot{\Omega}\dot{\phi}\cos\theta - \dot{\phi}\dot{\theta} \\[4pt] \Omega_3 = \dot{\Omega}\cos\theta + \dot{\phi} & \end{array}\right\}\qquad\text{(A4.3)}$$

The couples are the differentials

$$\left.\begin{array}{l}K_1 = -\partial V/\partial\theta \\[4pt] K_2 = -\partial V/\sin\theta\,\partial\Omega \\[4pt] K_3 = -\partial V/\partial\phi\end{array}\right\}\qquad\text{(A4.4)}$$

where V is the mutual potential of the Earth and the ring. V is independent of ϕ and so K_3 is zero. Hence Ω_3 is a constant, equal to n, the mean motion of the satellite in its orbit. n is much greater than the rate of change of θ or Υ and so the accelerations may be further simplified to

$$\left.\begin{array}{l}\dot{\Omega}_1 = n\dot{\Omega}\sin\theta \\[4pt] \dot{\Omega}_2 = -n\dot{\theta}\end{array}\right\}\qquad\text{(A4.5)}$$

Consider a point on the ring at angle ψ measured from N. The colatitude measured from the north pole of the Earth is χ, where

$$\cos\chi = \sin\theta\sin\psi$$

The potential corresponding to the equatorial bulge is

$$\frac{\mu}{r}\left(\frac{a}{r}\right)^2 J_2 \tfrac{1}{2}(3\cos^2\chi - 1)$$

and so the total mutual potential of the ring in the field of the Earth's bulge is

$$\frac{\mu}{r}\left(\frac{a}{r}\right)^2 J_2 \int_0^{2\pi} \tfrac{1}{2}(3\cos^2\chi - 1)\,\sigma r\,d\psi$$

where σ is the mass of the ring per unit length.

Hence

$$K_1 = -\frac{\partial V}{\partial\theta} = -\mu\left(\frac{a}{r}\right)^2 J_2\sigma \int_0^{2\pi} 3\sin^2\psi \sin\theta\cos\theta\,d\psi \qquad\text{(A4.6)}$$

that is

$$-\frac{3}{2}\frac{m}{r}\mu\left(\frac{a}{r}\right)^2 J_2 \sin\theta\cos\theta$$

since $m = 2\pi r\sigma$. V does not depend on Ω and so K_2 is zero.

Now make the following substitutions:

$$A = B = \tfrac{1}{2}mr^2, \quad C = mr^2$$
$$\dot{\phi} = n, \quad \dot{\Omega}_1 = n\dot{\Omega}\sin\theta$$
$$\Omega_2 = \dot{\Omega}\sin\theta, \quad \Omega_3 = n$$

The equation $A\dot{\Omega}_1 + (C - B)\Omega_2\Omega_3 = K_1$ then reads

$$mr^2 n\dot{\Omega}\sin\theta = -\frac{3}{2}\frac{m}{r}\mu\left(\frac{a}{r}\right)^2 J_2\sin\theta\cos\theta \qquad (A4.7)$$

or

$$\dot{\Omega} = -\frac{3}{2}n\left(\frac{a}{r}\right)^2 J_2\cos\theta \qquad (A4.8)$$

since $\mu = n^2 r^3$.

The result shows that the plane of the orbit rotates slowly about the polar axis of the Earth in the direction opposed to the motion of the satellite.

Modified Mercalli scale

  I. Not felt except by a few under especially favourable circumstances.
 II. Felt only by a few persons at rest, especially on upper floors of buildings. Delicately suspended objects may swing.
III. Felt quite noticeably indoors, especially on upper floors of buildings, but many people do not recognize it as an earthquake. Standing motor cars may rock slightly. Vibration like passing of lorry. Duration can be estimated.
 IV. Felt indoors by many during the day, outdoors by few. Some awakened at night. Dishes, windows, doors disturbed, walls make creaking sound. Sensation like heavy lorry striking building. Standing motor cars rocked noticeably.
  V. Felt by nearly everyone, many awakened. Some dishes, windows, etc., broken; a few instances of cracked plaster; unstable objects overturned. Disturbance of trees, poles, and other tall objects sometimes noticed. Pendulum clocks may stop.
 VI. Felt by all; many frightened and run outdoors. Some heavy furniture moved; a few instances of fallen plaster or damaged chimneys. Damage slight.
VII. Everybody runs outdoors. Damage negligible in buildings of good design and construction; slight to moderate in well-built ordinary structures; considerable in poorly built or badly designed structures; some chimneys broken. Noticed by people driving motor cars.
VIII. Damage slight in specially designed structures; considerable in ordinary substantial buildings, with partial collapse; great in poorly built structures. Panel walls thrown out of frame structures. Fall of chimneys, factory stacks, columns, monuments, walls. Heavy furniture overturned. Sand and mud ejected in small amounts. Changes in well water. Disturbs people driving motor cars.
 IX. Damage considerable in specially designed structures; well designed frame structures thrown out of plumb; great in substantial buildings, with partial collapse. Buildings shifted off foundations. Ground cracked conspicuously. Underground pipes broken.
  X. Some well-built wooden structures destroyed; most masonry and frame structures destroyed with foundations; ground badly cracked. Rails bent. Landslides considerable from river banks and steep slopes. Shifted sand and mud. Water splashes over banks.

XI. Few, if any, masonry structures remain standing. Bridges destroyed. Broad fissures in ground. Underground pipe lines out of service. Earth slumps and land slips in soft ground. Rails bent greatly.

XII. Damage total. Waves seen on ground surfaces. Lines of sight and level distorted. Objects thrown up into the air.

Equation of motion of the horizontal pendulum

The geometry of the horizontal pendulum is shown in figure 3.4. Let m be the mass, r the distance of its centre of mass from the hinge and I the moment of inertia mk^2, about the centre of mass. Let ϕ be the angle at which the hinge is inclined to the vertical and let θ be the angle by which the arm is displaced from its position of equilibrium. The equation of motion is readily derived from the Lagrangian form

$$\frac{\mathrm{d}}{\mathrm{d}t}\left(\frac{\partial L}{\partial \dot\theta}\right) = \frac{\partial L}{\partial \theta} \tag{A6.1}$$

where L, the Lagrangian, is equal to the difference between the kinetic energy, T, and the potential energy, V.

Now T is equal to the translational plus the rotational energy of the mass and arm. If the horizontal displacement of the ground in the direction perpendicular to the arm is u, the net velocity in that direction is $\dot u + r\dot\theta$. The angular velocity of the arm about the centre of mass is $\dot\theta$ and so

$$T = \tfrac{1}{2}m(\dot u + r\dot\theta)^2 + \tfrac{1}{2}I\dot\theta^2 \tag{A6.2}$$

When the arm is displaced by an angle θ from the position of equilibrium, the centre of mass is at a height $r(1 - \cos\theta)\sin\phi$ above its lowest point, and the potential energy is therefore

$$V = mgr(1 - \cos\theta)\sin\phi \tag{A6.3}$$

Thus

$$\frac{\partial T}{\partial \dot\theta} = mr(\dot u + r\dot\theta) + I\dot\theta$$

$$\frac{\partial V}{\partial \theta} = mgr\theta\sin\phi$$

taking θ to be small, and the equation of motion of the arm is

$$mr\ddot u + mr^2\ddot\theta + I\ddot\theta = -mgr\theta\sin\phi \tag{A6.4}$$

or

$$\ddot{\theta} + \omega^2\,\theta = -\alpha\ddot{u} \qquad\qquad \text{(A6.4a)}$$

where

$$\omega^2 = gr\sin\phi/(r^2 + k^2)$$

and

$$\alpha = r/(r^2 + k^2)$$

APPENDIX SEVEN

Elementary theory of elastic waves

Suppose that at a point in a solid with Cartesian coordinates (x, y, z) the solid is displaced from its undisturbed condition, and let the components of the displacement be (u, v, w). The components of normal strain are then

$$e_{xx}, e_{yy}, e_{zz} = \frac{\partial u}{\partial x}, \frac{\partial v}{\partial y}, \frac{\partial w}{\partial z}$$

while those of shear strain are

$$e_{xy} = \frac{\partial u}{\partial y} + \frac{\partial v}{\partial x}, \quad e_{yz} = \frac{\partial v}{\partial z} + \frac{\partial w}{\partial y}, \quad e_{zx} = \frac{\partial w}{\partial x} + \frac{\partial u}{\partial z}$$

The solid may also suffer a rotation with components

$$\xi_{xy} = \frac{\partial u}{\partial y} - \frac{\partial v}{\partial x}, \quad \xi_{yz} = \frac{\partial v}{\partial z} - \frac{\partial w}{\partial y}, \quad \xi_{zx} = \frac{\partial w}{\partial x} - \frac{\partial u}{\partial z}$$

The stress within a solid is specified by the stresses p_{xx}, p_{yy}, p_{zz} normal to the planes perpendicular to the Cartesian axes, and by the stresses p_{yz}, p_{zx}, p_{xy}, parallel to the coordinate planes.

A set of three mutually perpendicular axes may always be found such that the strains may be expressed as normal strains along the axes; similarly, a set of principal axes of stress can always be found. In an isotropic solid, in which the elastic properties do not depend on direction, the two sets of axes coincide. If they are taken as the (x, y, z) axes, the following linear relations hold for sufficiently small strains:

$$\left.\begin{aligned}
p_{xx} &= (\lambda + 2\mu)\, e_{xx} + \lambda e_{yy} + \lambda e_{zz} \\
p_{yy} &= \lambda e_{xx} + (\lambda + 2\mu)\, e_{yy} + \lambda e_{zz} \\
p_{zz} &= \lambda e_{xx} + \lambda e_{yy} + (\lambda + 2\mu)\, e_{zz}
\end{aligned}\right\} \tag{A7.1}$$

λ and μ are called the Lamé parameters for the solid. μ is the shear modulus.

Suppose that the solid is subject to a hydrostatic pressure. All the normal stresses are the same, equal to minus the pressure p. Hence

$$-3p = (3\lambda + 2\mu)(e_{xx} + e_{yy} + e_{zz}) \tag{A7.2}$$

$e_{xx} + e_{yy} + e_{zz}$ is called the *dilatation*, written as Δ. The ratio $-p/\Delta$ is the bulk modulus, or incompressibility, K, and so

$$K = \lambda + \tfrac{2}{3}\mu$$

If there are no body forces such as gravity, or if, as may usually be assumed for the Earth, the body forces are locally constant, then the balance between force and mass–acceleration leads to the following equations of motion in arbitrary Cartesian coordinates:

$$\left.\begin{aligned}
\rho\ddot{u} &= \frac{\partial p_{xx}}{\partial x} + \frac{\partial p_{yx}}{\partial y} + \frac{\partial p_{zx}}{\partial z} \\
\rho\ddot{v} &= \frac{\partial p_{xy}}{\partial x} + \frac{\partial p_{yy}}{\partial y} + \frac{\partial p_{zy}}{\partial z} \\
\rho\ddot{w} &= \frac{\partial p_{xz}}{\partial x} + \frac{\partial p_{yz}}{\partial y} + \frac{\partial p_{zz}}{\partial z}
\end{aligned}\right\} \qquad (A7.3)$$

Now it follows from the definitions of the strain components and from the relations between stress and strain for a linear isotropic solid, that

$$p_{xx} = \lambda\Delta + 2\mu e_{xx} = \lambda\Delta + 2\mu\,\partial u/\partial x$$

and

$$p_{yz} = \mu e_{yz} = \mu\left(\frac{\partial v}{\partial z} + \frac{\partial w}{\partial y}\right)$$

Hence

$$\rho\ddot{u} = (\lambda + \mu)\frac{\partial\Delta}{\partial x} + \mu\nabla^2 u \qquad (A7.4a)$$

$$\rho\ddot{v} = (\lambda + \mu)\frac{\partial\Delta}{\partial y} + \mu\nabla^2 v \qquad (A7.4b)$$

$$\rho\ddot{w} = (\lambda + \mu)\frac{\partial\Delta}{\partial z} + \mu\nabla^2 w \qquad (A7.4c)$$

These are the equations of motion on which the theory of wave propagation in an isotropic solid depends.

Now operate on equation (A7.4a) with $\partial/\partial x$, on (A7.4b) with $\partial/\partial y$ and on (A7.4c) with $\partial/\partial z$ and add the results:

$$\rho\left(\frac{\partial\ddot{u}}{\partial x} + \frac{\partial\ddot{v}}{\partial y} + \frac{\partial\ddot{w}}{\partial z}\right) = (\lambda + \mu)\left(\frac{\partial^2}{\partial x^2} + \frac{\partial^2}{\partial y^2} + \frac{\partial^2}{\partial z^2}\right)\Delta + \mu\nabla^2\left(\frac{\partial u}{\partial x} + \frac{\partial v}{\partial y} + \frac{\partial w}{\partial z}\right)$$

that is

$$\rho \frac{\partial^2 \Delta}{\partial t^2} = (\lambda + 2\mu) \nabla^2 \Delta \tag{A7.5}$$

Equation (A7.5), a form of the wave equation, shows that changes of dilatation will be propagated through the solid with a velocity α equal to

$$\{(\lambda + 2\mu)/\rho\}^{1/2} \quad \text{or} \quad \{(K + \tfrac{4}{3}\mu)/\rho\}^{1/2}$$

In such a wave there is no change of shape of any element of the solid, but just a change of volume.

Next, operate on (A7.4a) with $\partial/\partial y$ and on (A7.4b) with $\partial/\partial x$ and take the difference of the results:

$$\rho \frac{\partial^2 \xi_{xy}}{\partial t^2} = \mu \nabla^2 \xi_{xy} \tag{A7.6}$$

This again is a form of the wave equation and shows that a shear is propagated through the solid with a speed β equal to $(\mu/\rho)^{1/2}$. In such a wave there is no change of volume of an element of the solid.

Snell's law for elastic media

Consider two media in contact along the plane $z = 0$ as shown in figure 3.9. As in chapter 3, consider for simplicity a dilatational wave incident upon the boundary from above, making an angle θ_i with the normal to the boundary. Let its wave vector lie in the (x, z) plane so that quantities do not depend on y.

Since the displacements do not depend on y, the wave equations may be satisfied by writing

$$u = \frac{\partial \phi}{\partial x} + \frac{\partial \psi}{\partial z}, \quad w = \frac{\partial \phi}{\partial z} - \frac{\partial \psi}{\partial x} \tag{A8.1}$$

ϕ and ψ satisfy the equations

$$\ddot{\phi} = \alpha^2 \nabla^2 \phi \quad \text{(dilatational)}$$

$$\ddot{\psi} = \beta^2 \nabla^2 \psi \quad \text{(shear)} \tag{A8.2}$$

The incident dilatational wave in medium 1 will then be represented by

$$\phi = a_i \exp[-i\omega\{t - (x \sin \theta_i + z \cos \theta_i)/\alpha_1\}]$$

where ω is the angular speed, and the reflected wave by

$$\phi = a_r \exp[-i\omega\{t + (x \sin \theta_r + z \cos \theta_r)/\alpha_1\}]$$

where θ_r is the angle of reflection.

There may in addition be a reflected shear wave:

$$\psi = b_r \exp[-i\omega\{t + (x \sin \chi_r + z \cos \chi_r)/\beta_1\}] \tag{A8.3}$$

χ_r being an angle of reflection, while in medium 2 there will be transmitted dilatational and shear waves:

$$\left.\begin{array}{l} \phi = a_t \exp[-i\omega\{t - (x \sin \theta_t + z \cos \theta_t)/\alpha_2\}] \\ \psi = b_t \exp[-i\omega\{t - (x \sin \chi_t + z \cos \chi_t)/\beta_2\}] \end{array}\right\} \tag{A8.4}$$

The displacement u in the medium 1 is $\partial\phi/\partial x + \partial\phi/\partial z$ at $z = 0$ and is proportional to

$$\frac{a_i \sin \theta_i}{\alpha_1} \exp\left(\frac{i\omega x \sin \theta_i}{\alpha_1}\right) - \frac{a_r \sin \theta_r}{\alpha_1} \exp\left(\frac{i\omega x \sin \theta_r}{\alpha_1}\right)$$

$$- \frac{b_r \sin \chi_r}{\beta_1} \exp\left(\frac{i\omega x \sin \chi_r}{\beta_1}\right)$$

u in medium 2 is similarly

$$\frac{a_t \sin \theta_t}{\alpha_2} \exp\left(\frac{i\omega x \sin \theta_t}{\alpha_2} \right) + \frac{b_t \sin \chi_t}{\beta_2} \exp\left(\frac{i\omega x \sin \chi_t}{\beta_2} \right)$$

Now the displacements in the two media at $z = 0$ must be equal not only for all times but also all values of x and so the exponents in all the terms must be the same, that is

$$\frac{\sin \theta_i}{\alpha_1} = -\frac{\sin \theta_r}{\alpha_1} = -\frac{\sin \chi_r}{\beta_1} = \frac{\sin \theta_t}{\alpha_2} = \frac{\sin \chi_t}{\beta_2} \qquad \text{(A8.5)}$$

This is the generalized form of Snell's law for elastic media, and gives the angles of reflection and refraction for the different types of wave in terms of the respective velocities.

With this condition, the equation for the equality of displacements reads

$$a_i + a_r + b_r = a_t + b_t \qquad \text{(A8.6)}$$

a relation between the amplitudes of the waves. Three other (more complex) relations between the five amplitudes are obtained by substituting ϕ and ψ into the expressions for the displacement, w, and for the stresses on the common surface and requiring that they be continuous. The solution of these equations leads to values of a_r/a_i, b_r/a_i, a_t/a_i and b_r/a_i in terms of the angle of incidence and of the elastic moduli and densities of the two media. A similar calculation may be made for an incident shear wave.

Determination of seismic velocity within the Earth

The first step is to derive the general form of the relation between T and Δ. Consider a ray having the parameter p and originating on the surface at a point S. Let P be any point on the ray such that OP is equal to r (O is the centre of the Earth) and let the angle POS be θ. Let the arc length SP be s. Then the angle i that the ray makes with the radius OP at P is given by

$$\sin i = r\, d\theta/ds \qquad (A9.1)$$

Since $p = (r\sin i)/v$ it follows then

$$p = \frac{r^2}{v}\frac{d\theta}{ds}$$

Now

$$ds^2 = dr^2 + r^2\, d\theta^2$$

and so

$$\left(\frac{dr}{d\theta}\right)^2 = \left(\frac{r^2}{pv}\right)^2 - r^2$$

or

$$\frac{d\theta}{dr} = \pm\frac{p}{r(\eta^2 - p^2)^{1/2}} \qquad (A9.2)$$

where $\eta = r/v$.

Thus p is the value of the variable η at the lowest point of the ray.

Now $\int d\theta = \Delta$, and so on integrating from the deepest point of the ray (at which $r = r_p$, say) up to the surface ($r = r_0$) it follows that

$$\Delta = 2p \int_{r_p}^{r_0} \frac{dr}{r(\eta^2 - p^2)^{1/2}} \qquad (A9.3)$$

Alternatively,

$$\left(\frac{dr}{ds}\right)^2 = 1 - r^2\left(\frac{d\theta}{ds}\right)^2 = 1 - \frac{p^2 v^2}{r^2}.$$

and, since $T = \int \mathrm{d}t = \int \mathrm{d}s/v$, the integral being taken along the ray,

$$T = \int \frac{1}{v}\frac{\mathrm{d}s}{\mathrm{d}r}\,\mathrm{d}r$$

Now

$$\frac{\mathrm{d}s}{\mathrm{d}r} = r\,\frac{r}{v(\eta^2 - p^2)^{1/2}}$$

and so

$$T = 2\int_{r_p}^{r_0} \frac{\eta^2\,\mathrm{d}r}{r(\eta^2 - p^2)^{1/2}} \tag{A9.4}$$

It follows from the expressions for T and $\varDelta$ that

$$T = p\,\varDelta + 2\int_{r_p}^{r_0} \frac{(\eta^2 - p^2)^{1/2}\,\mathrm{d}r}{r} \tag{A9.5}$$

The expression for $\varDelta$ may be written in the equivalent form

$$\varDelta = 2\int_{p}^{\eta_0} \frac{p}{r(\eta^2 - p^2)^{1/2}}\,\frac{\mathrm{d}r}{\mathrm{d}\eta}\,\mathrm{d}\eta \tag{A9.6}$$

Let us now see how to determine v as a function of r from equation (A9.6), given $\varDelta$ as a function of T. Let r' be a value of r between r_0 and the lowest point of the ray for which r takes the value r_p. Consider a set of rays for which p covers the range from η_0, the surface value, to η', the value of p at r'. Multiply both sides of equation (A9.6) by $(p^2 - \eta'^2)^{-1/2}$ and integrate with respect to p:

$$\int_{\eta'}^{\eta_0} \frac{\varDelta\,\mathrm{d}p}{(p^2 - \eta'^2)^{1/2}} = 2\int_{\eta'}^{\eta_0} \mathrm{d}p \int_{p}^{\eta_0} \frac{\mathrm{d}r}{\mathrm{d}\eta}\,\frac{p\,\mathrm{d}\eta}{r(\eta^2 - p^2)^{1/2}(p^2 - \eta'^2)^{1/2}}$$

It is permissible to change the order of integration:

$$\int_{\eta'}^{\eta_0} \frac{\varDelta\,\mathrm{d}p}{(p^2 - \eta'^2)^{1/2}} = 2\int_{\eta'}^{\eta_0} \mathrm{d}\eta \int_{p}^{\eta_0} \mathrm{d}p\,\frac{\mathrm{d}r}{\mathrm{d}\eta}\,\frac{p}{r(\eta^2 - p^2)^{1/2}(p^2 - \eta'^2)^{1/2}}$$

On integrating by parts, the left side becomes

$$\left[\varDelta\,\mathrm{arccosh}\left(\frac{p}{\eta'}\right)\right]_{\eta'}^{\eta_0} - \int_{\eta'}^{\eta_0} \frac{\mathrm{d}\varDelta}{\mathrm{d}p}\,\mathrm{arc\,cosh}\left(\frac{p}{\eta'}\right)\mathrm{d}p$$

which equals

$$\int\limits_{0}^{\Delta'} \operatorname{arccosh}\left(\frac{p}{\eta'}\right) \mathrm{d}\Delta$$

the first term being zero because Δ is zero when $p = \eta_0$, while arc cosh(p/η') is zero when $p = \eta'$.

Thus

$$\int\limits_{0}^{\Delta'} \operatorname{arccosh}\left(\frac{p}{\eta'}\right) \mathrm{d}\Delta = \pi \ln\left(\frac{r_0}{r'}\right) \qquad (A9.4)$$

The times of travel of waves from place to place over the surface of the Earth give T as a function of Δ, and from them p, which is equal to $\partial T/\partial \Delta$, may also be obtained as a function of Δ. A value of η' is selected and the integral on the left side evaluated—suppose the result is $I'(\eta')$. Then

$$r' = r_0/\exp(I'/\pi)$$

that is, r' is obtained as a function of η' when the calculations are repeated for other values of η'. Since $\eta' = r'/v'$, v' is obtained as a function of r', that is, the distribution of v with radius is derived.

Surface waves

Let the top surface of discontinuity (figure A10.1) be taken as the plane $z = 0$ (z positive downwards) and take perpendicular axes of x and y in the surface. Let a disturbance be propagated in the direction of x, so that all quantities are constant with respect to y. Then the equations of motion, which are

$$\left.\begin{aligned}
\rho\ddot{u} &= (\lambda + \mu)\frac{\partial \Delta}{\partial x} + \mu\nabla^2 u \\[2mm]
\rho\ddot{v} &= (\lambda + \mu)\frac{\partial \Delta}{\partial y} + \mu\nabla^2 v \\[2mm]
\rho\ddot{w} &= (\lambda + \mu)\frac{\partial \Delta}{\partial z} + \mu\nabla^2 w
\end{aligned}\right\} \tag{A10.1}$$

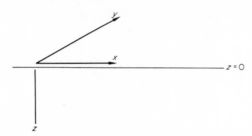

Figure A.10.1 Semi-infinite solid

can be satisfied by setting

$$u = \frac{\partial \phi}{\partial x} + \frac{\partial \psi}{\partial z}$$

and

$$w = \frac{\partial \phi}{\partial z} - \frac{\partial \psi}{\partial x}$$

where ϕ, ψ and v are all functions of x and z only.

Then

$$\left.\begin{aligned}
\nabla^2\phi &= \frac{\partial u}{\partial x} + \frac{\partial w}{\partial z} = \nabla \\[2mm]
\nabla^2\psi &= \frac{\partial u}{\partial z} - \frac{\partial w}{\partial x}
\end{aligned}\right\} \tag{A10.2}$$

and the equation of motion may be written as

$$\left.\begin{aligned}
\ddot{\phi} &= \alpha^2 \nabla^2 \phi \\
\ddot{\psi} &= \beta^2 \nabla^2 \phi \\
\ddot{v} &= \beta^2 \nabla^2 v
\end{aligned}\right\} \tag{A10.3}$$

For a harmonic wave propagated in the x direction independent of y and decaying exponentially with z, ϕ, ψ and v will all be proportional to

$$\exp\{-\gamma z + i(\kappa x - vt)\}$$

Let

$$\phi, \psi, v = (\phi_0, \psi_0, v_0) \exp\{-\gamma z + i(\kappa x - vt)\}$$

Then

$$\ddot{\phi}, \ddot{\psi}, \ddot{v} = -v^2\, e^{\chi'}(\phi_0, \psi_0, v_0)$$

where

$$\chi' = -\gamma z + i(\kappa x - vt)$$

and so

$$\nabla^2(\phi, \psi, v) = (\gamma^2 - \kappa^2)\, e^{\chi'}(\phi_0, \psi_0, v_0)$$

The wave equation therefore implies

for ϕ: $\qquad\qquad\qquad -v^2 = \alpha^2(\gamma^2 - \kappa^2)$

for ψ and v: $\qquad\qquad -v^2 = \beta^2(\gamma^2 - \kappa^2)$

whence

$$\gamma = \pm \kappa \left(1 - \frac{c^2}{\alpha^2}\right)^{1/2} \quad \text{or} \quad \pm \kappa \left(1 - \frac{c^2}{\beta^2}\right)^{1/2} \tag{A10.4}$$

respectively; $c = v/\kappa$ is the wave velocity.

Now γ must be real and positive for the disturbance to die away from $z = 0$, so that c must be less than β (which is less than α).

Write

$$\kappa \left(1 - \frac{c^2}{\alpha^2}\right)^{1/2} = \gamma_p$$

$$\kappa \left(1 - \frac{c^2}{\beta^2}\right)^{1/2} = \gamma_s$$

Then

$$u = (i\kappa\phi_0 e^{-\gamma_p z} - \gamma_s \psi_0 e^{-\gamma_s z}) e^{i\chi}$$
$$v = v_0 e^{-\gamma_s z} e^{i\chi} \left.\vphantom{\begin{array}{c}1\\1\\1\end{array}}\right\} \quad \text{(A10.5)}$$
$$w = (-\gamma_p \phi_0 e^{-\gamma_p z} - i\kappa\psi_0 e^{-\gamma_s z}) e^{i\chi}$$

where

$$\chi = \kappa x - vt$$

The forms (A10.5), corresponding to a mixed dilatational and shear disturbance, satisfy the equations of motion, are independent of y, and vanish when z is large. It is now necessary to satisfy the boundary conditions at $z = 0$, which are that there should be no stress component perpendicular to the surface, namely

$$p_{zz} = 0, \quad p_{xz} = 0$$

Now

$$p_{xz} = \mu \left(\frac{\partial u}{\partial z} + \frac{\partial w}{\partial x} \right)$$

and so is proportional to

$$-2i\kappa\gamma_p \phi_0 e^{-\gamma_p z} + (\gamma_s^2 + \kappa^2) \psi_0 e^{-\gamma_s z}$$

then, at $z = 0$,

$$-2i\kappa\gamma_p \phi_0 + (\gamma_s^2 + \kappa^2) \psi_0 = 0$$

Also

$$p_{zz} = \lambda \frac{\partial u}{\partial x} + (\lambda + 2\mu) \frac{\partial w}{\partial z} \quad \text{since} \quad \frac{\partial v}{\partial y} = 0$$

and so, at $z = 0$,

$$\{(\lambda + 2\mu) \gamma_p^2 - \lambda\kappa^2\} \phi_0 - 2i\kappa\mu\gamma_s \psi_0 = 0$$

The two boundary conditions are of the form

$$A\phi_0 + B\psi_0 = 0$$
$$C\phi_0 + D\psi_0 = 0 \left.\vphantom{\begin{array}{c}1\\1\end{array}}\right\} \quad \text{(A10.6)}$$

and have non-zero solutions only if

$$AD - BC = 0$$

Accordingly

$$(\gamma_s^2 + \kappa^2) \{(\lambda + 2\mu) \gamma_p^2 - \lambda\kappa^2\} - 4\mu\kappa^2 \gamma_p \gamma_s = 0 \quad \text{(A10.7)}$$

Substitute for γ_s and γ_p and divide by κ^2:

$$\left(2 - \frac{c^2}{\beta^2}\right)\left\{2 - \frac{\lambda + 2\mu}{\mu}\frac{c^2}{\alpha^2}\right\} = 4\left(1 - \frac{c^2}{\alpha^2}\right)^{1/2}\left(1 - \frac{c^2}{\beta^2}\right)^{1/2}$$

or since

$$\frac{\lambda + 2\mu}{\mu} = \frac{\alpha^2}{\beta^2}$$

$$\left(2 - \frac{c^2}{\beta^2}\right)^2 = 4\left(1 - \frac{c^2}{\alpha^2}\right)^{1/2}\left(1 - \frac{c^2}{\beta^2}\right)^{1/2} \tag{A10.8}$$

(A 10.8) is the equation to determine the velocity of the waves. In general, it must be solved numerically but a good idea of the character of Rayleigh waves can be obtained by setting $\lambda = \mu$, a condition nearly satisfied for many actual materials, including rocks.

Then $\alpha^2 = 3\beta^2$ and

$$\frac{c^6}{\beta^6} - 8\frac{c^4}{\beta^4} + \frac{56}{3}\frac{c^2}{\beta^2} - \frac{32}{3} = 0$$

An inadmissible solution is $c = 2\beta$ (c must be less than β) and the other solutions satisfy

$$3\frac{c^4}{\beta^4} - 12\frac{c^2}{\beta^2} + 8 = 0$$

or

$$c^2 = 2\beta^2(1 \pm 3^{-1/2})$$

For c less than β,

$$c = (0 \cdot 8)^{1/2}\beta$$

$$= 0 \cdot 919\beta$$

Then

$$\gamma_p = 0 \cdot 848\kappa, \ \gamma_s = 0 \cdot 393\kappa$$

and

$$u = (e^{-0 \cdot 848\kappa z} - 0 \cdot 578\,e^{-0 \cdot 393\kappa z})\,e^{ix}$$

$$w = (0 \cdot 848\,e^{-0 \cdot 848\gamma z} - 1 \cdot 468\,e^{-0 \cdot 393\kappa z})\,e^{ix}$$

There remains one boundary condition to satisfy, namely $p_{yz} = 0$, that is

$$\frac{\partial v}{\partial z} + \frac{\partial w}{\partial y} = 0$$

or

$$\frac{\partial v}{\partial z} = 0$$

since all quantities are independent of y.

Thus

$$\gamma_s^2 v_0 e^{-\gamma_s z} e^{i\chi} = 0$$

which can only be satisfied if v_0 is zero.

In Rayleigh waves, therefore, the disturbance lies in the plane perpendicular to the free surface and containing the direction of propagation, and there is no motion in the free surface perpendicular to the direction of propagation. A given point in the material moves in an ellipse, since $|u|$ is not equal to $|w|$ and the form of the ellipse changes with depth since the ratio $|u|/|w|$ changes with depth.

The equation for the velocity of Rayleigh waves depends only on the properties of the material and so, as for body waves, the velocity is independent of period.

Now consider the possibility of a surface disturbance with the displacement in the y direction, perpendicular to the direction of propagation. Such a disturbance has been seen to be impossible on a uniform semi-infinite medium, so consider a layer of different properties lying upon the top of the semi-infinite medium. Take the interface between the two layers to be the plane $z = 0$ and let the thickness of the overlying layer be H (figure A10.2). Call the upper layer 1 and the lower, semi-infinite layer, 2. Since the disturbance in 1 does not necessarily decrease with z it may be written as

$$v = (v_1 e^{-\gamma_1 z} + v_1' e^{\gamma_1 z}) e^{i\chi} \qquad (\chi = \kappa x - vt)$$

while in medium 2

$$v = v_2 e^{-\gamma_2 z} e^{i\chi}$$

where

$$\gamma_1 = \kappa (1 - c^2/\beta_1^2)^{1/2}$$

but is not restricted to be real, while

$$\gamma_2 = \kappa (1 - c^2/\beta_2^2)^{1/2}$$

and must be real.

The following boundary conditions must now be satisfied.
(1) At $z = 0$, the displacements are the same in the two layers:

$$v_1 + v_1' = v_2$$

(2) At $z = 0$, the shear stresses p_{xy} and p_{yz} must be the same in the two layers, that is $\partial v/\partial x$ and $\partial v/\partial z$ are the same. Continuity of $\partial v/\partial x$ is equivalent to continuity of v, while continuity of $\partial v/\partial z$ gives

$$\mu_1(v_1 - v_1')\gamma_1 = \mu_2 v_2 \gamma_2$$

(3) At $z = H$, the shear stress p must vanish, that is $\partial v/\partial z = 0$ at $z = H$, or

$$v_1 \exp(-\gamma_1 H) = v_1' \exp(\gamma_1 H)$$

There are now three equations to be satisfied by v_1, v_1' and v_2; written in matrix form they are

$$\begin{bmatrix} 1 & 1 & -1 \\ \mu_1 \gamma_1 & -\mu_1 \gamma_1 & -\mu_2 \gamma_2 \\ \exp(-\gamma_1 H) & -\exp(\gamma_1 H) & 0 \end{bmatrix} \begin{bmatrix} v_1 \\ v_1' \\ v_2 \end{bmatrix} = 0 \qquad \text{(A10.9)}$$

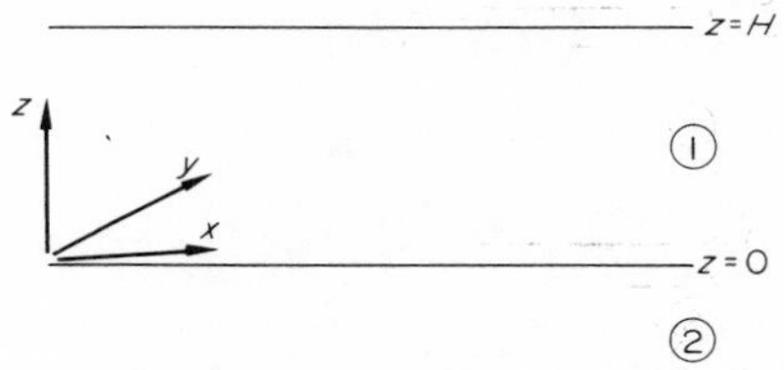

Figure A.10.2 Semi-infinite solid with surface layer

and, for a non-zero solution, the determinant of the coefficients must vanish:

$$\mu_1\gamma_1(e^{-\gamma_1 H} - e^{\gamma_1 H}) + \mu_2 \gamma_2(e^{-\gamma_1 H} + e^{\gamma_1 H}) = 0$$

or

$$\mu_2 \gamma_2 = \mu_1 \gamma_1 \tanh \gamma_1 H \qquad \text{(A10.10)}$$

For this equation to be satisfied, given that γ_1 is imaginary, γ_2 must be real. It follows from the relation

$$\gamma = \pm \kappa \left(1 - \frac{c^2}{\beta^2}\right)^{1/2}$$

that β_1 must be less than β_2 and that c must lie between them.
The equation for c may be written

$$\mu_1\left(1 - \frac{c^2}{\beta_1^2}\right)^{1/2} - \mu_2\left(1 - \frac{c^2}{\beta_2^2}\right)^{1/2} \tan\left\{\kappa H \left(\frac{c^2}{\beta_2^2} - 1\right)^{1/2}\right\} = 0 \qquad \text{(A10.11)}$$

from which it is evident that c, the wave velocity, depends on κ and so on the period of the waves.

APPENDIX ELEVEN

Precession

The geometry is shown in figure A11.1. E is the pole of the ecliptic and P is the pole of the equator and the angle EOP is the inclination, θ, of the equator to the ecliptic. OP is the axis of rotation of the Earth. The planes of the equator and ecliptic intersect in the line of nodes, NN′, the direction of which makes the angle Ω with the direction of the first point of Aries (♈). OX is any direction fixed in the equator and makes an angle ϕ with ON; ϕ increases at the rate $\tilde{\omega}$, the spin angular velocity of the Earth. The position of the Moon (or the Sun) in the ecliptic is measured by the angle λ between the direction of the Moon or the Sun and the

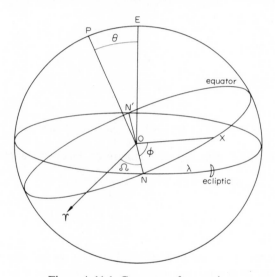

Figure A.11.1 Geometry of precession

line of nodes. The angles θ, Ω, ϕ are Eulerian angles (Landau and Lifshitz, 1969) for the specification of the rotating Earth in the fixed reference system of the ecliptic.

The direction OP is a principal axis of inertia of the Earth and the moment of inertia about it is denoted by C (see chapter 2). In dealing with the precession of the Earth it is sufficient to take the moment of inertia about any axis in the equator to be the same, namely A. The motion of the Earth is found by equating the rates of change of the components of the angular momentum of the Earth to the respective couples and, since the directions OP and any two perpendicular directions in the equatorial plane are principal directions, it is simplest to take components of angular momentum with respect to these axes. Let the components of angular velocity be Ω_1 about the direction OP, Ω_2 about the direction OX and Ω_3 about the direction OY, perpendicular to OX and OP. Then, as in appendix 4,

$$\left.\begin{aligned}
\Omega_1 &= \dot{\Omega}\cos\theta + \dot{\phi} \\
\Omega_2 &= \dot{\Omega}\sin\theta\sin\phi + \dot{\theta}\cos\phi \\
\Omega_3 &= \dot{\Omega}\sin\theta\cos\phi - \dot{\theta}\sin\phi
\end{aligned}\right\} \tag{A11.1}$$

(Landau and Lifshitz, 1969).

Now, as for the problem of the ring satellite treated in appendix 4, the direction OX is arbitrary in the Earth because all axes in the equator are taken to be equivalent. The expressions for the angular velocities may therefore be simplified by taking OX to be instantaneously in the direction ON so that ϕ is zero. With that convention,

$$\left.\begin{aligned}
\Omega_2 &= \dot{\theta}; \quad \dot{\Omega}_2 = \dot{\Omega}\dot{\phi}\sin\theta + \ddot{\theta} \\
\Omega_3 &= \dot{\Omega}\sin\theta; \quad \dot{\Omega}_3 = -\dot{\theta}\dot{\phi} + \frac{\mathrm{d}}{\mathrm{d}t}(\dot{\Omega}\sin\theta)
\end{aligned}\right\} \tag{A11.2}$$

Let the equations of motion be written in Euler's form (see appendix 4):

$$\left.\begin{aligned}
C\dot{\Omega}_1 &= K_3 \\
A\dot{\Omega}_2 + (C - A)\Omega_1\Omega_3 &= K_2 \\
A\dot{\Omega}_3 + (A - C)\Omega_1\Omega_2 &= K_3
\end{aligned}\right\} \tag{A11.3}$$

where K_1, K_2, K_3 are the couples about the respective axes.

The couples are those arising from the attraction of the Moon or the Sun upon the Earth's equatorial bulge and so are respectively

$$\left.\begin{aligned}
K_1 &= -M'\,\partial V/\partial\phi \\
K_2 &= -M'\,\partial V/\partial\theta \\
K_3 &= -M'\,\partial V/\sin\theta\,\partial\dot{\Omega}
\end{aligned}\right\} \tag{A11.4}$$

where V is the potential of the Earth at the Moon or the Sun and M' is the mass of the Moon or the Sun. The $-\mu/r$ term in the potential does not depend on angle (ignoring the eccentricities of the orbits of the Sun and the Moon and other small effects) and the only term of significance in the potential is

$$\frac{G(C - A)}{r^3}\tfrac{1}{2}(3\cos^2\chi - 1)$$

where r is the distance of the Sun or the Moon and χ is the colatitude of the Sun or

Moon measured from the north pole of the Earth. $\cos \chi$ must be expresssed in terms of the Eulerian angles and the longitude of the Moon measured from the first point of Aries:

$$\cos \chi = \sin(\lambda - \Omega) \sin \theta$$

It follows that V is independent of ϕ (on account of the postulated symmetry of the Earth about the polar axis) and so $K_1 = 0$. Accordingly Ω_1 is a constant, in fact, the spin angular velocity of the Earth, $\tilde{\omega}$. Hence

$$\dot{\Omega} \cos \theta + \dot{\phi} = \tilde{\omega} \tag{A11.5}$$

Now $\tilde{\omega}$ is so much greater than all other angular velocities that as for the ring satellite the latter may everywhere be neglected in comparison with $\tilde{\omega}$ and, in particular, $\dot{\phi}$ may be replaced by $\tilde{\omega}$. The remaining two equations of motion then reduce to

$$\left.\begin{array}{l} \dot{\Omega} C\tilde{\omega} \sin \theta = -M' \, \partial V/\partial\theta \\[2mm] -\dot{\theta}C\tilde{\omega} = M' \, \partial V/\sin \theta \, \partial\Omega \end{array}\right\} \tag{A11.6}$$

Thus

$$\left.\begin{array}{l} \dot{\Omega} = -\dfrac{GM'}{\tilde{\omega}r^3} \dfrac{C-A}{C} \, 3 \sin^2(\lambda - \Omega) \cos \theta \\[4mm] \dot{\theta} = -\dfrac{GM'}{\tilde{\omega}r^3} \dfrac{C-A}{C} \tfrac{3}{2} \sin 2(\lambda - \Omega) \sin \theta \end{array}\right\} \tag{A11.7}$$

Let the factor

$$\frac{GM'}{\tilde{\omega}r^3} \frac{C-A}{C}$$

be denoted by k, so that

$$\dot{\Omega} = -\tfrac{3}{2}k\{1 - \cos 2(\lambda - \Omega)\} \cos \theta$$

and

$$\dot{\theta} = -\tfrac{3}{2}k \sin 2(\lambda - \Omega) \sin \theta$$

θ varies very little in practice so that it may be replaced by its mean value, θ_0, on the right side. λ is equal to nt, where n is the orbital angular velocity of the Sun or the Moon.

Thus

$$\left.\begin{array}{l} \Omega = -\tfrac{3}{2}kt \cos \theta_0 + \dfrac{3}{4n} \, k \cos \theta_0 \sin 2(nt - \Omega) \\[4mm] \theta = \dfrac{3}{4n} \, k \sin \theta_0 \cos 2(nt - \Omega) \end{array}\right\} \tag{A11.8}$$

where θ_0 has the value $23°.27'$.

Ω changes at the steady rate $-\frac{3}{2}k\cos\theta_0$ on the average—this is the *precession of the equinoxes* for it gives rise to a steady movement of the line of nodes about the pole of the ecliptic. Superposed on this steady motion, there is an oscillatory motion with argument $2(nt - \Omega)$. Similarly, θ oscillates with the same argument and the two oscillations make up the nutation.

APPENDIX TWELVE

Some biographical notes

Airy, Sir George Biddell 1801–92 (English)

Astronomer, Astronomer Royal. Educated at Cambridge, where he was successively Lucasian Professor and Plumian Professor of Astronomy. As Astronomer Royal, he reorganized the magnetic observatory at Greenwich. In 1854 he used pendulums to measure the change of gravity between the top and bottom of a mine in South Shields and so estimated the mean density of the Earth. He suggested the form of isostatic compensation that goes by his name, whereby the light crust is thicker under mountains than under the oceans.

Bouguer, Pierre 1698–1758 (French)

Mathematician and Professor of Hydrography at Croisic, Brittany, where he was born. Studied navigation and the magnetic variation and made pioneer investigations in photometry. Took part with de la Condamine in the expedition to Peru that established the polar flattening of the Earth. Published in 1749 *Figure de la Terre Determinée*. Observed the horizontal attraction of M. Chimborazo on a plumb bob.

Clairaut, Alexis-Claude 1713–65 (French)

Mathematician, who took part in the expedition to Lapland (1736) that demonstrated that the Earth was flattened at the poles, in confirmation of Newton's ideas. Was the first to connect the shape of the surface of a rotating body with the variation of gravity over it, a result known as 'Clairaut's theorem' (*Théorie de la Figure de la Terre*, 1743). He also studied the orbit of the Moon (*Théorie de la Lune*, 1750).

Darwin, Sir George Howard 1845–1912 (English)

Eldest son of Charles Darwin. Mathematician, Professor of Astronomy at Cambridge. He extended the theory of the gravity field of the Earth, studied its elastic yielding under tidal forces and made extensive investigations of the tides and of the motion of the Moon.

Gauss, Karl Friedrich 1777–1855 (German)

Mathematician, Director of the Observatory of Göttingen from 1807, where, in association with Weber, he erected a magnetic observatory. He devised a number of magnetometers and introduced the representation of the surface magnetic field by spherical harmonics. He worked extensively in celestial mechanics and elaborated the method of least squares.

Gilbert, William 1544–1603 (English)

Born at Colchester. Fellow of St. John's College, Cambridge and M.D., travelled extensively in Europe and practised as physician in London. Appointed physician to Queen Elizabeth (1599) and reappointed to James I. Published in 1600 *De Magnete Magneticisque Corporibus, et de Magno Magnete Tellure*, an account of his experiments on magnetized bodies and electrical forces and of his idea that the Earth is a large magnet, so giving the first consistent account of the magnetic field of the Earth.

Halley, Edmund 1656–1742 (English)

Astronomer, born in London, educated at Oxford. On account of the scientific expeditions he undertook from the age of 19 onwards, he did not complete the requirements for the M.A. degree and it was conferred on him (1768) by the instruction of Charles II. Subsequently Savilian Professor of Geometry. He was elected to the Royal Society at the age of 22 and subsequently acted as Secretary. He was Astronomer Royal from 1730. Apart from astronomical work of outstanding importance he was the first to observe the magnetic variation and suggested that the Earth's field originated in a central core that could rotate relatively to the outer shell. He suggested a connection between the aurora borealis and the magnetic field. In the course of astronomical work on St Helena, he detected the change of rate of his pendulum clock in consequence of the difference of gravity between London and St Helena. Perhaps his greatest contribution to science was persuading and assisting Newton to publish the *Principia*, which was done at his expense.

Kater, Henry 1777–1835 (English)

Soldier and geodesist. Best known for devising the 'reversible pendulum' for the absolute measurement of the acceleration due to gravity His measurements of differences of gravity within the British Isles were of extraordinary precision for his day.

Kelvin, William Thompson, Lord 1824–1907 (Scottish)

The most distinguished and versatile British physicist of the 19th century. Professor of Natural Philosophy at Glasgow. He made fundamental advances in thermodynamics and devised the electric telegraph. He made a famous and controversial estimate of the age of the Earth based on its rate of cooling.

Maskelyne, Nevil 1732–1811 (English)

Astronomer Royal. In 1772 he proposed, and in 1774 effected, a determination of the horizontal attraction of Schiehallion, an isolated mountain near Loch Tay in Perthshire, from which Charles Hutton calculated the mean specific gravity of the Earth to be 4·5.

Newton, Sir Isaac 1642–1727 (English)

Newton's work of course transcends geophysics but, in the narrow sense, he showed that a rotating Earth should be flattened at the poles and that Halley's observations of the rate of the pendulum clock at St Helena supported the conclusions.

Pratt, John H. d. 1871 (English)

Archdeacon of Calcutta. Mathematician and author of books on mechanics and the figure of the Earth. As a result of his studies of deflections of the vertical in the Himalayas, he suggested that the density of the crust below mountains was less than below oceans—the 'Pratt' form of isostatic compensation. A monument to him is set up in Calcutta Cathedral.

Rayleigh, John William Strutt, Lord (1842–1919) (English)

Mathematician and physicist, Chancellor of Cambridge University. Showed that elastic waves should be guided by the free surface of a solid. His work *The Theory of Sound* contains the fundamentals of much of classical physics. 'Rayleigh's principle' is extensively used in the study of oscillating systems. He succeeded Maxwell as Cavendish Professor of Experimental Physics at Cambridge and was instrumental in the discovery of argon.

Rutherford, Ernest, Lord 1871–1937 (New Zealand)

Rutherford, the founder of nuclear physics and Cavendish Professor of Physics at Cambridge, first pointed out in a lecture at Yale in 1906 that the ages of rocks could be found from the decay products of radioactive elements within them. The first estimates made use of the accumulation of helium.

Stokes, Sir George Gabriel (1819–1903) (Irish)

Mathematician, Master of Pembroke College, Cambridge. His greatest work was in optics and hydrodynamics but in geophysics he is known for 'Stokes's theorem' on the determination of the form of the geoid from the values of gravity upon it.

Turner, H. H. 1861–1930 (English)

Professor of Astronomy at Oxford. He, together with Zöppritz, compiled the first reliable tables of the times of travel of seismic pulses through the Earth.

Index